Circling the Square

How the Conquest Altered the Shape of Time in Mesoamerica

Circling the Square

How the Conquest Altered the Shape of Time in Mesoamerica

ANTHONY AVENI

American Philosophical Society
Philadelphia • 2012

Transactions of the
American Philosophical Society
Held at Philadelphia
For Promoting Useful Knowledge
Volume 102, Part 5

ISBN: 978-1-60618-025-9

US ISSN: 0065-9746

Library of Congress Cataloging-in-Publication Data

Aveni, Anthony F.
  Circling the square : how the conquest altered the shape of time in
Mesoamerica / Anthony F. Aveni.
     p. cm. — (Transactions of the American Philosophical Society,
ISSN 0065-9746 ; v. 102, pt. 5) Includes bibliographical references and
index.
  ISBN 978-1-60618-025-9
  1. Calendar—Mexico. 2. Maya calendar. 3. Mexico—History—
Conquest, 1519-1540. 4. Mexico—History—Spanish colony, 1540-1810.
5. Central America—History—To 1821. I. Title. II. Series: Transactions
of the American Philosophical Society ; v. 102, pt. 5.
  F1219.3.C2A94 2012
  972'.02—dc23
                              2012041205

Contents

Preface   vii
Acknowledgments   ix
List of Illustrations   xi

I. INTRODUCTION   1

II: TIME'S CYCLE, TIME'S CIRCLE   9

III: CALENDAR WHEELS FROM CENTRAL MEXICO   17
   Motolinía's Calendar Wheel and Its Variants   19
   The Calendar Wheels of Sahagún   41
   Codex Mexicanus Calendar Wheels   47
   The Duran and Tovar Calendar Wheels   49
   The Boban Calendar Wheel   50
   Resumé and Discussion   53

IV. MAYA CALENDAR WHEELS   57
   Landa's Katun Wheel   59
   Calendar Wheel in the Book of Chilam Balam of Kaua   61
   Calendar Wheel in the Book of Chilam Balam of Chumayel   63
   Calendar Wheels in the *Book of Chilam Balam of Ixil*   64
   Resumé and Discussion   68

V. INDIGENOUS TIME IN THE SQUARE   71

VI. SUMMARY   85

NOTES   91

REFERENCES   101

Index   *111*
Color insert begins following page   56

Preface

Since antiquity, time in the West has been represented in circular form, the gear wheels of time churning out duration in endless years. At the time of the Spanish conquest, dials on round clock faces looked down from facades of cathedrals and town halls, their sonorous tones chiming out the hours that directed people's religious and workaday lives. To judge from the way Spanish chroniclers describe it, the circle was also the principal mode of temporal expression among the New World natives they sought to Catholicize. Because the Gregorian Calendar Reform coincided with the proliferation of the clock, the subject of time was very much on the minds of sixteenth-century scholars.

This study, which examines an array of calendar circles appearing in manuscripts from Central Mexico and the Maya area of Yucatan from the time of sixteenth-century contact up to the eighteenth century, demonstrates that this was decidedly not the case. Rather, the indigenous quadripartite way of perceiving space rendered the expression of time in the form of a square or rectangle.

This study follows the gradual intrusion of Western calendrical particulars into the native format, such as the correspondence between the months and the phases of the moon, the base-twelve format of the divisions of the year and the zodiac, the wind compass, the Olympiad, the concept of the Jubilee, and the leap year.

We acquire insight regarding the tension in the first generation of native scribes after the conquest, who were working with radically different ways of knowing, between the imposed requirement to change the way they thought about time and the innate desire to preserve their heritage and their identity.

Acknowledgments

I am especially indebted to Eleanor Wake, who brought various versions of Central Mexican documents analyzed herein to my attention and who, via a number of lengthy discussions and correspondences, generously conveyed her vast knowledge and informed opinions about many of them to me. She also kindly consented to doing translations where needed. Susan Milbrath generously offered detailed comments on a series of drafts, and also helped with translations. Victoria Bricker contributed to my work in this same vein, especially with regard to the Maya documents. Christine Hernández was very helpful with discussions regarding Aubin ms. 20. I thank Elizabeth Boone for commenting on an earlier draft, and Michel Oudijk for the same; he also brought the Boturini Collection to my attention. Additional thanks are owed to Guilhem Olivier and Scott Sessions, who shared information on inventories of Hispanic texts employed by the chroniclers. I am also grateful to Barbara Fash who, several years ago, showed me the calendar wheels housed in the Harvard Peabody Museum collection, which originally inspired my investigation. I thank my assistant, Diane Fanney, for the valuable day-to-day work involved in putting my presentations together. Lastly, I also thank David Carrasco and the Mesoamerican Archive at Harvard University for allowing me to present my early ideas on this research.

Illustrations

Figure 1.1   A popular cogged-wheel representation of   3
              Maya time in the contemporary literature. The
              13- and 20-day cycles that represent the *tzol-
              kin*, or sacred round of 260 days, roll around
              the larger seasonal wheel of 365 days.

Figure 1.2   A pair of cogged calendar wheels depicted in   5
              the colonial Maya *Book of Chilam Balam of Ixil*,
              a Maya town.

Figure 1.3   Calendar Wheels mesh in Codex Mexicanus,   6
              p. 9.

Figure 1.4   Stone embedded in a church wall in Central   7
              Mexico.

Figure 2.1   Examples of Medieval and Renaissance wheel-
              shaped diagrams known as *rotae*.

Figure 2.1a: One example of Medieval and Renaissance   12
              wheel-shaped diagrams known as *rotae*. Coper-
              nican diagram of the solar system showing
              planets on circular orbs surrounded by the
              quintessential constellations of the zodiac.

Figure 2.1b: Another example of a Medieval and Renais-   14
               sance wheel-shaped diagram know as *rotae*. A
               Wheel of Time for Common Folk. Each time
               frame bears a pictorial representation of the
               appropriate activity.

Figure 2.1c: A third example of Medieval and Renaissance   15
               wheel-shaped diagrams know as *rotae*. The
               Winds (ca. 1200 A.D.).

Figure 3.1   Calendar Wheel of Motolinía, 1903.   20

Figure 3.2   European Wheel of solar declinations for suc-   21
               cessive dates of the seasonal year, 1550.

Figure 3.3   Calendar wheel in the Codex Tlaxcala known   24
               as Muñoz Camargo Wheel No. 1, (1583–5).

Figure 3.4   Valadés Calendar Wheel (1579).   27

Figure 3.5   Gemelli-Careri Wheel, 1697.   29

Figure 3.6   Representation of a coiled snake swallowing its 31
               tail from a Spanish Reportorio.

Figure 3.7.   Muñoz Camargo Wheel No 2. Note the crease   33
               showing where it had been folded.

Figure 3.8.   Veytia Wheel No. 5.   34

Figure 3.9   Lorenzana Wheel (1770).   35

Figure 3.10   Possible reproduction of the Veytia calendar   37
               wheel #5. Manuscript on parchment.

Figure 3.11   Veytia Wheel No. 7.   39

Figure 3.12   Veytia Wheel No. 1.   40

Figure 3.13   Serna Wheel No. 2.   42

Figure 3.14   Sahagún Calendar Wheel.   43

Figure 3.15   Durán Calendar Wheel, *Codice Durán*.   51

Figure 3.16   Boban Calendar Wheel.   52

Figure 4.1   Landa's Katun Wheel.   60

Figure 4.2   Wind Compass in the Book of Chilam Balam   62
              of Kaua.

Figure 4.3   Calendar Wheel in Book of Chilam Balam of   63
              Kaua.

Figure 4.4   Chilam Balam of Chumayel Wheel.   65

Figure 4.5   Translation of Chumayel Wheel.   66

Figure 5.1   Diagram showing spiral course of the day-   75
              name sequence on page 25 of the Codex
              Borgia.

Figure 5.2   Codex Fejérváry-Mayer, p. 1. Codices Selecti, 76
              xxvi.

Figure 5.3   Codex Madrid.   77

Figure 5.4   Mendoza Codex, p. 1.   79

Figure 5.5   Veytia Calendar "Wheel" No. 3.   80

Figure 5.6   Aubin 1.   81

Figure 5.7   Aubin ms. 20.   82

Figure 5.8   Pecked cross circles, early examples of pre-   83
              Columbian time in the round.

Figure 5.9   Pecked cross Teotihuacan No. 2 (cf. Figures   84
              5.2 and 5.3).

I

Introduction

> And he gave the universe the figure which is proper and natural
> . . . .Wherefore he turned it, as in a lathe, round and spherical,
> with its extremities equidistant in all directions from the centre,
> the figure of all figures most perfect . . .
>
> —Plato, *Timaeus*

The Maya calendar is commonly represented in the literature in the form of interlocking cogged wheels. One popular rendition of this scheme that appeared in *National Geographic* magazine carried the following caption: "The wheels of time ground exceeding fine for the Maya" (Figure 1.1) (Stuart 1975, p. 783). This graphic portrayal of the commensuration of smaller calendrical cycles with larger ones in the form of gears meshing together resembles what one finds in the innards of a mechanical clock. The Maya time-machine metaphor also appeared much earlier in a widely circulated monograph on Maya astronomy by J.E. Teeple (1930, p. 39); there the 260-day sacred count, or *tzolkin*, is presented as a rotating toothed wheel.

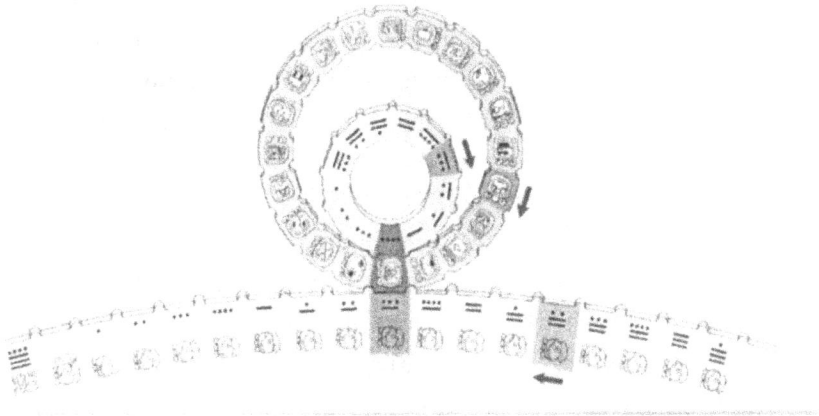

Figure 1.1 A popular cogged-wheel representation of Maya time in the contemporary literature. The 13- and 20-day cycles that represent the *tzolkin*, or sacred round of 260 days, roll around the larger seasonal wheel of 365 days.

Reprinted from G. Stuart. (1975). "The Maya Riddle of the Glyphs." *National Geographic* 148(6): 783, with permission.

But Maya calendrical wheels that resemble clockwork extend much further back in history; for example, on page 39 of the colonial Maya *Book of Chilam Balam of Ixil* (Figure 1.2). In Central Mexico a similar representation can be found on p. 9 of the postcontact Codex Mexicanus (Figure 1.3; **see also color plate opposite p. 56**). Additionally, a pair of tangent wheels, one bearing 13 and the other 9 floral symbols, likely intended to display the cycling of the coefficients of the sacred count (there called the *tonalpohualli*) with the Nine Lords of the Night, appears on a carved stone embedded in the wall of the sixteenth-century church of San Matias Tlalancaleca, Puebla (Noguera 1964; Garcia Cook 1973, pp. 25–29) (Figure 1.4). These calendar wheel representations might lead one to believe that indigenous Mesoamerican time was conceived in the minds of those who created it as circular, perhaps even derived from some sort of indigenous mechanistic view of nature. But these conclusions are far from the truth.

The purpose of this study is to examine the historical underpinnings of the sea change in the pictorial representation of calendrical time that took place in Hispanic Mesoamerica. We begin, necessarily, with an exploration of the roots of the deeply entrenched circular view of time in the West expressed in the epigraph. Next we examine, in chronological order, transformations in the representation of several examples of colonial period calendar wheels, beginning in Central Mexico then moving to the Maya region. This inquiry leads to a contrast with indigenous representations of time, which were conveyed largely in square or quadrangular rather than circular form; this is the subject of the final portion of the narrative. Thus, the postcontact expression of Mesoamerican time emerges as one of many examples of the alteration in representation of fundamental native concepts affected by the clash of cultures we call the Spanish conquest.

Though early postcontact native scribes were likely knowledgeable about the form and content of pre-Columbian codices (Edmonson 1974, p. 7; Tozzer 1941, p. 28, n. 156), there were many other documents extant that influenced the way they participated in the process of representing time in the early colonial period texts in which many calendar

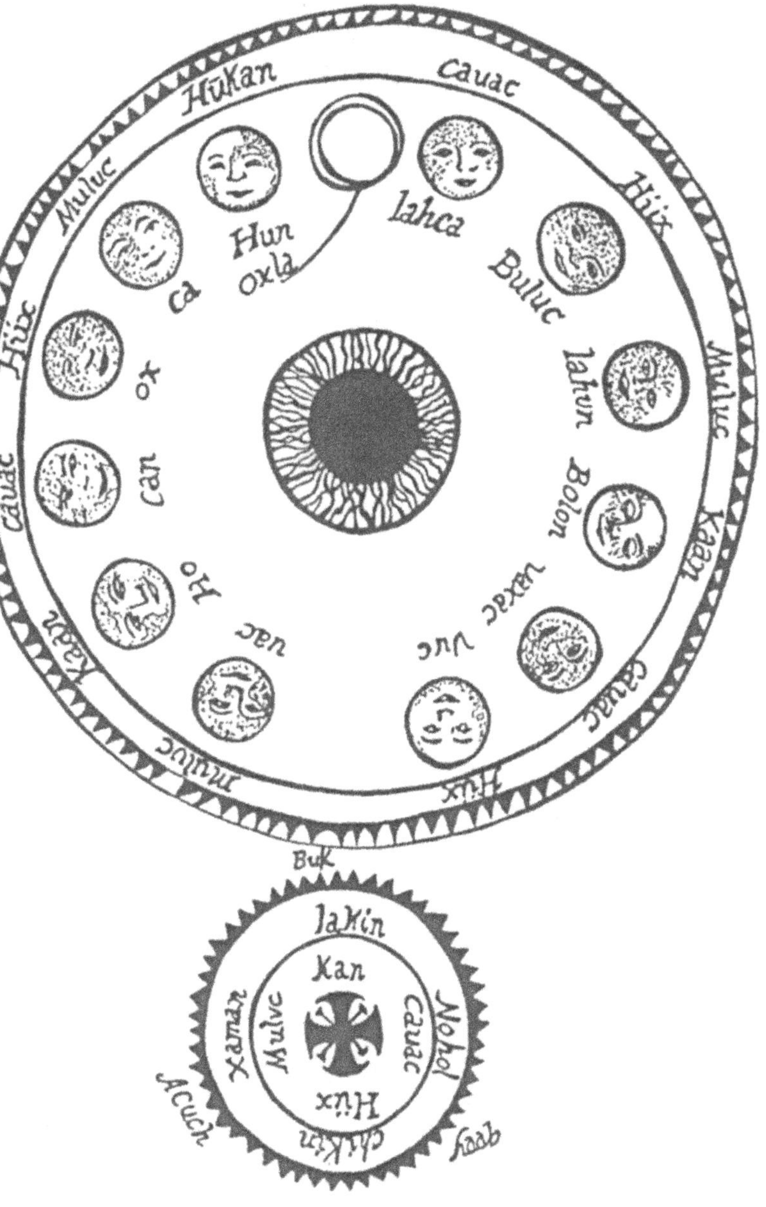

Figure 1.2 A pair of cogged calendar wheels depicted in the colonial Maya *Book of Chilam Balam of Ixil*, a Maya town.

Reprinted from V. Bricker and H. Miram. (2002). Pub 68, fig 39, with permission of the Middle American Research Institute.

Figure 1.3   Calendar Wheels mesh in Codex Mexicanus, p. 9.

Codex Mexicanus, p. 9, from Central Mexico. Reprinted with permission of the Bibliothéque Nationale de France (BnE) (http://amoxcalli.org.mx/laminas.php?id=023-024&ord_lamina=023_09&act=con)

wheels appear. A major issue in this study will be to determine exactly who designed and produced the wheels, especially the early ones. Were they Spaniards, or natives inspired by European cosmographical and astronomical/astrological texts known as *reportorios*?

Given what we have learned from the early Spanish chroniclers, one can imagine a native scribe assisting one of them in trying to make sense of two vastly different systems of time reckoning, one inherited, the other forcibly thrust upon him. Imagine, too, how puzzled he must have been when forced by the inquisitor priest to answer questions relating to the meaning of symbols and numbers in old pre-Columbian documents and failing to offer up the anticipated answers. Wrote one European intruder:

> In this [Maya] town another calendar was found . . . but I do not understand it. The ones who make them are the same ones who are the superstitious curers. It is almost impossible to find them because of the . . . great secrecy they keep about this. This [sin] is asked about very frequently in confession, but they all deny it. (Weeks et al. 2009, p. 79)

Figure 1.4   Stone embedded in a church wall in Central Mexico.

Courtesy of E. Wake. (2010). *Framing the Sacred: The Indian Churches of Colonial Mexico*. Norman: University of Oklahoma Press, fig. 5.6b. The image is redrawn from Doris Heyden, *Mitología y simbolismo de la flora en el México prehispánico*, Mexico City: UNAM/IIA, 1983.

And another:

> These Ajq'ijes [calendar keepers], who are like men who have a pact with
> the devil, not only mislead the other Indians to believe and hold infallible
> the prognostication and omen but also make them preserve idolatry and
> make them worship [the idols] as deceivers by transforming into nahuals
> in the form of ferocious animals to frighten them. (Weeks et al., p. 27)

Little wonder that following the burning of more than 99 percent
of the Maya codices shortly after contact, native calendar keepers felt
the need to create clandestine documents, such as the *Books of Chilam
Balam* and the *Annals of the Cakchiquels*. How else could they preserve
their sacred heritage and maintain their indigenous religious identity?
In some instances scholars have been able to discover which European
documents were accessible to informants. For example, Bricker and
Miram (2002) have demonstrated that a number of passages in the
Maya Book of Chilam Balam of Kaua were not only strongly influenced
by but also, in some instances, copied directly from European reportorios such as those of Chavez (1581) and Zamorano (1585). What Bricker
and Miram have shown for the Maya, Spitler (2005a) and Oudijk and
Castaneda de la Paz (2010) have made evident for the Central Mexican
colonial calendars. Spitler reviews several reportorios known to have
been in the libraries of Motolinía and Sahagún, which employed Nahua
chronologists who may have acquired some of these European books,[1]
whereas Oudijk and Castaneda have established a number of direct
connections between colonial calendar wheels and European
documents.

II

Time's Cycle, Time's Circle

FOR THE GREEKS, TO WHOM WE ATTRIBUTE a large share of the development of Western thought, the circle, as the epigraph preceding chapter 1 suggests, was divine. Following the idea of Plato's creator-craftsman, who gave the world its sacred shape, Aristotle employed the sphere, or rotated circle, as the basis of his theory of natural place. The idealized form of the universe he envisioned consisted of a set of nested spheres housing the hierarchically arranged five elements, from bottom to top: Earth, Water, Air, Fire, and Quintessence ("fifth essence"). Eudoxus, Aristotle's fourth-century-B.C. pupil, devised a *simulacrum*, or geometric model, of the Earth-centered solar system that consisted of a series of rotating concentric spheres capable of predicting future positions of the sun, moon, planets, and stars, all denizens of the Quintessence.

This spherical model of celestial motion proved capable of generating precise positions that matched reality, thus "saving the appearances," as the Greeks termed the lofty goal that underpinned the creation of these mechanically conceived models, that is, until it was replaced by a more efficient model, also based on circular orbits, devised by the Alexandrian astronomer Ptolemy (ca. A.D. 150). Ptolemy's model placed the planets on small circles, termed *epicycles*, the centers of which moved along bigger circles, called *deferents*, the latter centered on the Earth. With minor adjustments, the better to save the appearances, Ptolemy's version of a circular cosmos would suffice for more than a millennium, until more precise observational data acquired during the scientific Renaissance rendered it, too, relatively inaccurate for predicting planetary positions.

The Copernican revolution, which dates from the 1543 publication of the seminal work *De Revolutionibus Orbium Coelestium*, set the Earth, along with the other planets, in motion about the sun—again on circular orbits (cf. Figure 2.1a; see also color plate opposite p. 56). The Platonic tradition, revivified by the first printing of his work in 1484 (Kristeller 1978), thus remained alive and well. The unquestioned dominance of the "circular dogma," as Arthur Koestler (1959, p. 57) has called it, served as the basic form underlying all representations of celestial

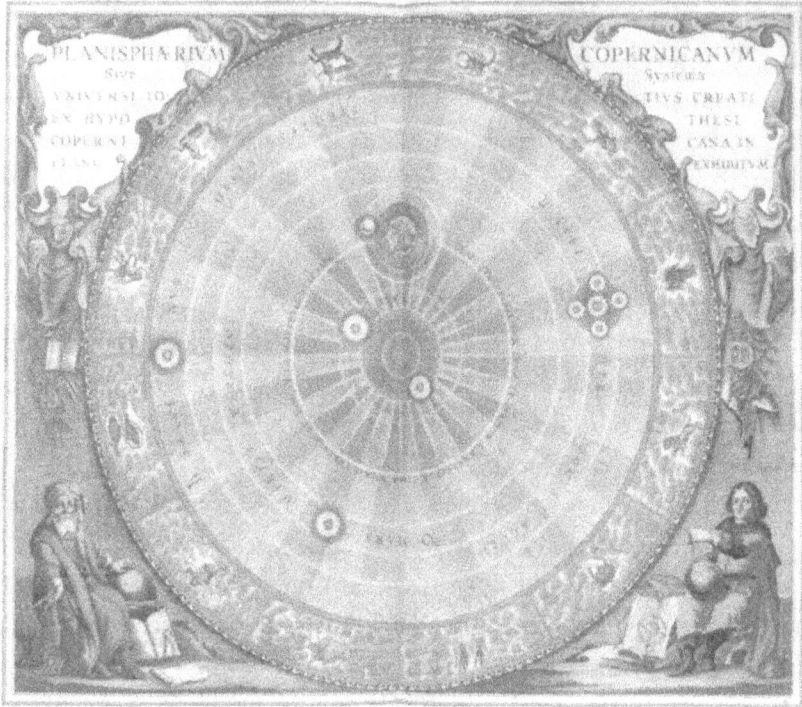

Figure 2.1a   One example of Medieval and Renaissance wheel-shaped diagrams known as *rotae*. Copernican diagram of the solar system showing planets on circular orbs surrounded by the quintessential constellations of the zodiac.

Reprinted with permission of Musée Stewart, Montreal.
(http://loveforlife.com.au/files/ResizedPlanisphere_of_Copernicus_jk.jpg)

motion in the West until the early seventeenth century, when Johannes Kepler, exasperated by his continued failure to fit the extant observational data relating to the orbit of the planet Mars to a satisfactory circular model, was forced to the conclusion that planetary orbits are not circular, but rather elliptical (Beer and Beer 1975).

In the two millennia that separated Plato and Kepler, explanations of just about every natural phenomenon in the West have relied on the circular form. Circular diagrams known as *rotae* (wheels) began to appear widely in books written in the Middle Ages and earlier. Among

the most famous works on *rotae* was Isidore of Seville's (A.D. 560–636)
De Natura Rerum, often called *Liber Rotarum* because of the profundity
of wheel-shaped diagrams displayed in it (cf. Murdock 1984, Kline
2001). Isidore's influence spread to other handbook authors, including
Macrobius and the Venerable Bede. Wheels of the Months delineated
lists of the months of the year in radial sectors, indicating the number
of days in each and the date in the Roman kalends of the first of each
month. Seasonal wheels, such as that of Bartholomaeus Anglicus (1485),
also segmented in circular compartments, pictured the zodiacal constellations, often listing the qualities of summer (hot, dry) opposite winter
(cold, wet), and so on. Pictures showing human activities conducted
during various times of the seasonal cycle are included, for example,
planting in April, thrashing hay in October, and so on (e.g., Figure
2.1b) There were specialized wheels for computing the date of Easter,
the rising and setting points of the sun, and the number of hours of
moonshine. Wind compasses, often depicting faces with open mouths
blowing air, listed qualities assigned to various winds (Figure 2.1c).

The subject matter dealt with in Medieval and Renaissance *rotae*
could range well beyond natural phenomena; for example, wheels of
consanguinity diagrammed relations among kin, offering definitions of
various ancestors and descendants, thus allowing the user to trace lines
of filiation. Other wheels diagrammed the reduction of syllogisms, the
Ages of Man (with poetic accounts of each), and the taxonomy of Old
Testament patriarchs. By the fifteenth century any learned European
would have been quite familiar with a vast diversity of taxa expressed
in circular form.

Kepler's discovery of elliptical orbs notwithstanding, today the
doctrine of the circle thrives. In common parlance the words *cycle* and
circle are practically synonymous. The *circle* is defined as a closed curve
on which every point is equidistant from the center. Compare this
definition with that of the word *cycle*: "a period within which a round
of regularly occurring events is completed."[2] Indeed the Greek word
kyklus, from which our word "cycle" derives, is defined as a circle, or
that which combines the form of a circle or wheel. We commonly

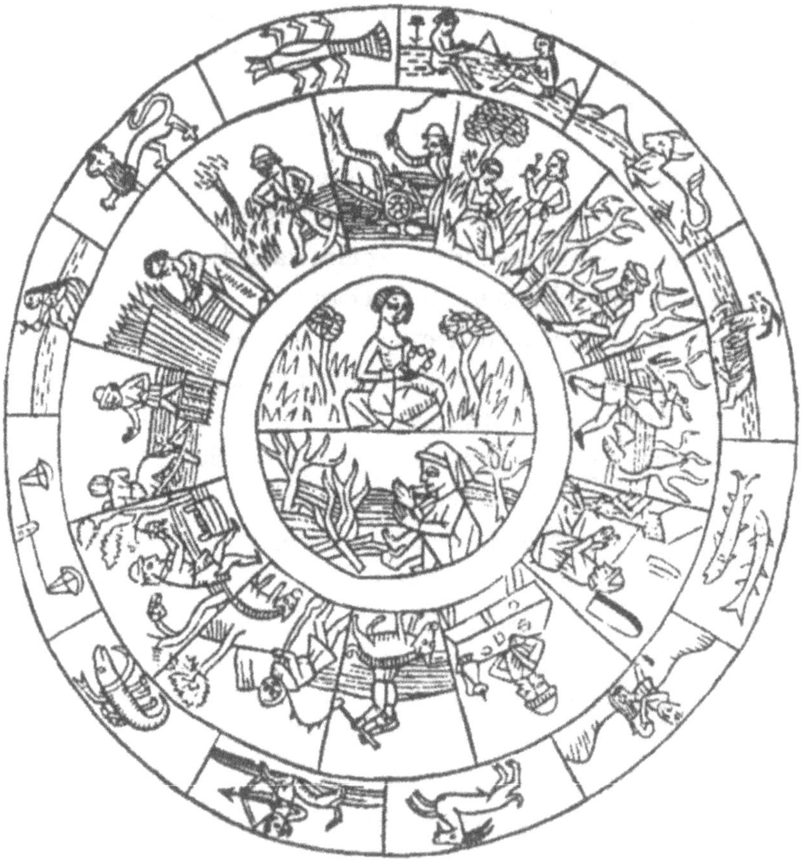

Figure 2.1b Another example of Medieval and Renaissance wheel-shaped diagrams known as *rotae*. A Wheel of Time for Common Folk. Each time frame bears a pictorial representation of the appropriate activity.

Reprinted from Bartholomaeus Anglicus. (1485). *De Proprietatibus*. Reproduced by permission of The Huntington Library, San Marino, California.

differentiate between cyclic time, for example, the round of the days of the week, or the seasonal round, and linear time in the Judeo-Christian historical tradition, which has been compared to a series of events, each occurring but once along a continuum that spans the interval between creation and the end of the world. Clearly, the con-

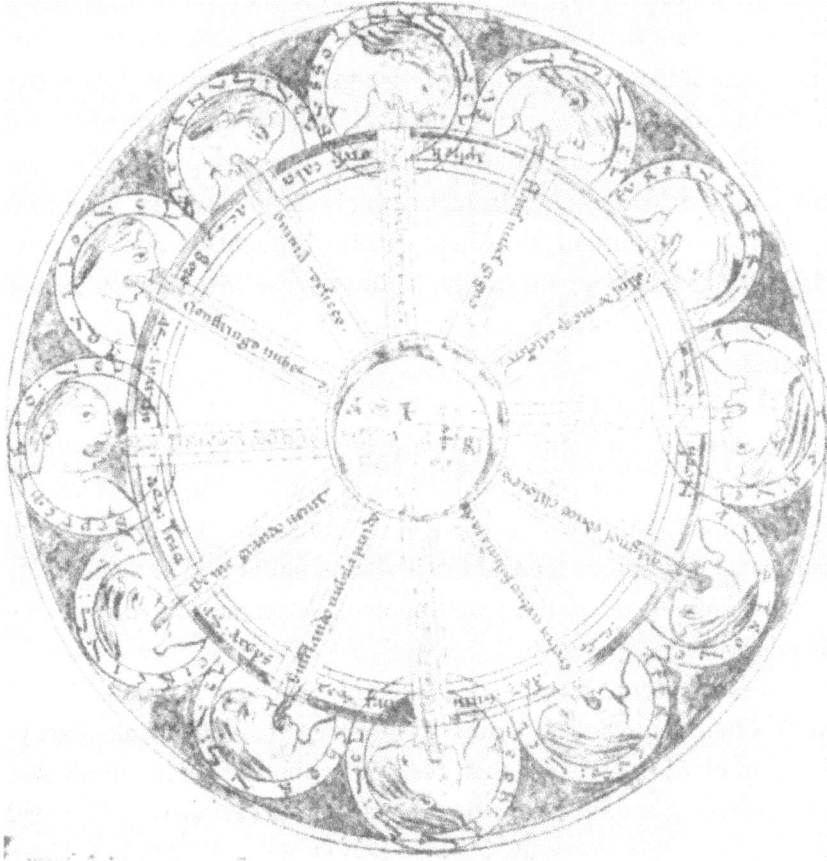

Figure 2.1c A third example of Medieval and Renaissance wheel-shaped diagrams known as *rotae*. The Winds (ca. 1200 A.D.).

Reprinted with permission from the Walters Art Gallery, Baltimore, MS W. 73, f.iv.

trasting geometrical models for each are the circle and the straight line, respectively.[3]

Interestingly, the beginning of the Spanish conquest coincides with two seminal developments in the Western concept of understanding time. First, the conquest of the New World and the dissemination of the Copernican world view happened during the same generation. First-edition copies of *De Revolutionibus* (1543), now housed in the Biblioteca

Nacional in Madrid and the Biblioteca de San Lorenzo (El Escorial), arrived in Spain from northern Europe as early as 1545. Philip II purchased one of these volumes himself (Gingerich 2002, p. 199). Theories of the organization of the solar system were major topics of conversation in intellectual circles. To judge by the representations of space and time that appear in the early chronicles, geocentrism was the dominant theory. More important, the shape of time brought over to the New World in the minds of those who would inquire into native customs for the purpose of conversion would have been subject to the rule of the circle.

The second development, the realization that time, by then represented in every mercantile European city by the round face of the chiming clock at the center of town, was out of joint, would lead to the Gregorian reform of the calendar. Completed in 1582, the reform had been a widespread topic of discussion in public and private sectors at home and abroad in the years following Hispanic contact with the New World.

In early colonial documents of this era, both from highland Mexico and Yucatán, there appear a number of representations of calendars in the form of wheels. As we shall see, the content of these wheels was often informed and their construction sometimes executed by educated natives, for whose use some of them may have been intended. These artifacts offer revealing details about the process of communicating, teaching, and learning basic concepts of great contemporary importance across two very diverse cultures, one curious about exploring idolatrous native concepts to assist in conversion, the other struggling to cling to a threatened cultural heritage, while at the same time also looking for ways to become active agents in the making of their own history. Drawing on the relatively few previous studies mentioned earlier, first of the calendar wheels from Central Mexico and then of those from the Maya area, we next explore a selection of calendar wheels that best demonstrates how time appears to have been apprehended in the dialog between conqueror and conquered. We begin with the earliest examples.

III

Calendar Wheels From Central Mexico

Motolinía's Calendar Wheel and Its Variants

Historian Alexis de Tocqueville once likened history to a gallery of pictures in which there are few originals and many copies. Indeed there are several versions of what would appear to be the earliest colonial representation of indigenous calendars in the round and they are interesting to examine because they reveal changes through time. The prototype in the set appears in *Motolinía's History of the Indians of New Spain* (Figure 3.1). Steck (1951, p. 53) speculates that Motolinía may have acquired the idea of trying to explain the structure of the Aztec calendar in the form of a circular diagram by abstracting it from the shape of the Aztec sunstone, though there is no proof he ever saw this artifact.[4] Oudijk and Castañeda de la Paz (2010, p. 120) have noted the stark resemblance the Motolinía wheel bears to a figure denoting solar declinations that appears in a 1550 Hispanic text on nautical astronomy by Pedro de Medina, which may have been its prototype (Figure 3.2 [see also color plate opposite p. 56]; Cuesta Domingo 1998).

In the accompanying text Motolinía refers to two concentric wheels, the outer representing the 52-year cycle, which commensurates the cycles of 260 and 365 days, and the inner cycle of 20-day names of the former (*tonalpohualli*). Thus, the outer ring, the reading of which proceeds clockwise, is divided into 52 compartments. Following a Latin cross, which appears at the 12 o'clock position, are four pictures: a house symbol and the words "*principio xihuitl*," or "year count start," a picture of a flint and "2. *tecpatl* [flint] *xihuitl*"; a picture of a house and "3. *calli* [house] *xihuitl*"; and finally "4. *tochtli* [rabbit] *xihuitl*." In each case the words are written along the outermost segment of the compartment that contains the picture. The fifth and sixth compartments read "5. *acatl xihuitl*" and "6. *tecpatl xihuitl*," respectively; however, these are not accompanied by pictures.[5] The remaining 46 outer compartments are left blank.

Moving inward, we find a series of what appear at first glance to be 13 concentric rings divided into 20 numbered sectors, beginning at approximately the "12:30" position on the innermost ring, 1, 2,

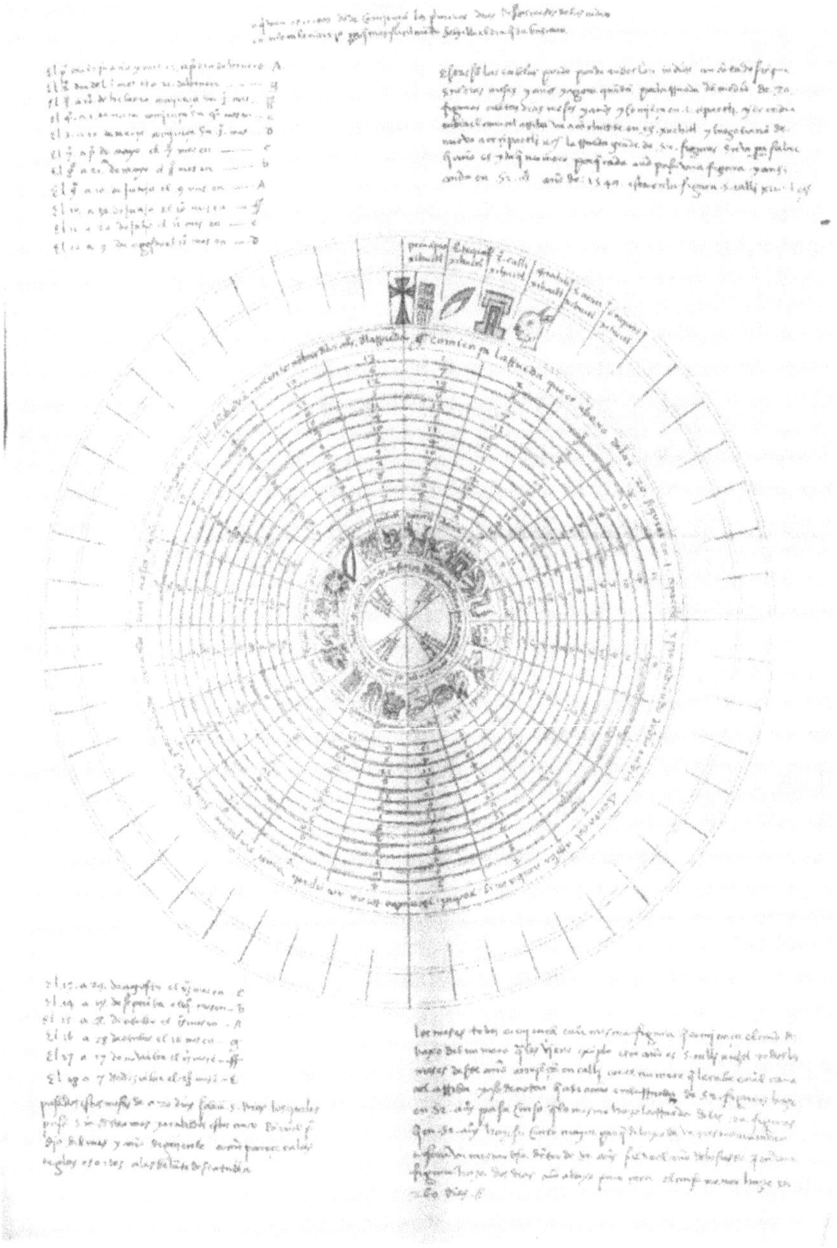

Figure 3.1 Calendar Wheel of Motolinía, 1903.

Reprinted with permission from G. Pimentel, ed. 1903. *Memoriales de Toribio de Motolinía*. Mexico: En Casa del Editor.

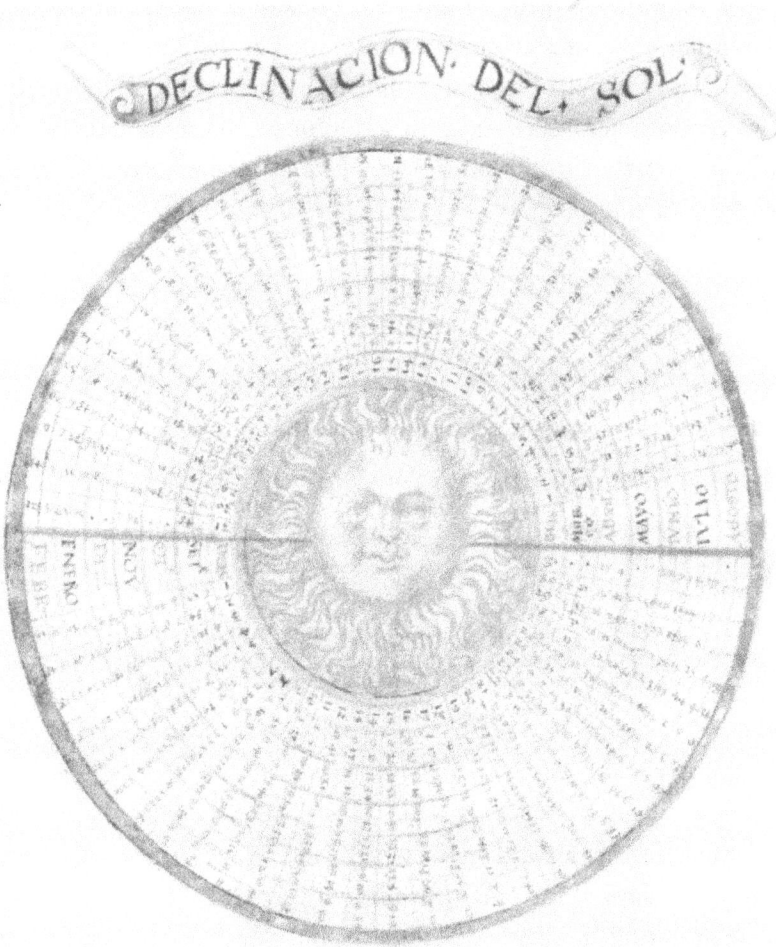

Figure 3.2 European Wheel of solar declinations for successive dates of the seasonal year, 1550.

3, . . . 13, and running clockwise, whereafter the count resumes. But closer inspection reveals that the series of concentric circles resolves to a roughly improvised, tightly wound spiral, which removes the necessity of jumping discontinuously to the next outer ring on completing a revolution so as to read the calendar.[6] Whether this spiral structure, which mimics that in precontact codices, such as the Borgia (see Figure 5.1, p. 75) is deliberate on the chronicler's part is not evident. The final (260th) entry places the user at the 11:30 position on the inner wheel, which is separated from the 52 year-bearer segments by a circular inscription that reads:

comiença la rrueda que es abaxo de.las.20.figuras

en.1 cipactli y proçediendo de día en dia el caracol arriba acaba

en.13. xochitl.y de nuevo torna a ce çipactli va así sín parar

siempre va[n] procediendo contando dias meses y años sín nínguna

excesio[n] Ariba va[n] escritos los no[m]bres de los años de la

rrueda. [The wheel beneath the 20 figures starts on 1 cipactli and

proceeding day by day the conch (spiral) reaches the top on 13

xochitl and returns again to 1 cipactli continuing in this way

without stopping always moving on to count days months and

years with no (exceptions?). Above, the names of the wheel's years

are written.][7]

Moving further inward on the wheel, we find the names and pictorials for all 20 days of the *tonalpohualli*, with Cipactli (crocodile) at the top, and rotating in sequence to the "Olin," or Motion, sign, which appears as a sun face. Xochitl, represented by a flower, completes the 20-day circuit. A second inscription positioned on the next inner ring reads:

Exta rrueda siempre xamas comjensa . . . el año en una

de las 4 rosas de la+in(t/f)erior. [This wheel always begins the year

on one of the four flowers in the interior.][8]

Finally, eight lines segmented into pairs form triangles (the "flowers" in the text) and are arranged in the form of an X, radiating outward from the center. These little triangles, labeled by year-bearers *calli*, *tochtli, acatl,* and *tecpatl,* are connected to the pictorials on the inner ring.

Motolinía's wheel further attempts to connect the 52-year round to Western historical time. The circular drawing is accompanied by a written text on how to "find the year, month, and day in which we are" (Steck 1951, p. 55). In the list of rules that appears alongside the wheel describing its component parts, Motolinía demonstrates how to accomplish this for the year 1549, which evidently corresponds to the approximate time during which he was working on his manuscript. Thus he tells users where to look on the wheel to find the equivalent place in the Aztec calendar for the dates 1 Jan., 10 Feb., . . . 30 June, 20 July, 9 Aug. . . . 7 Dec of that year. He also includes an aid to determining leap year equivalents.[9]

Motolinía's interest seems to focus exclusively on the mechanics of time. There is no reference to time's meaning in any indigenous context. With seeming disregard of the association between native divinatory ritual and the various periodicities he delineates in his calendar wheel, Motolinía rather intends the device to be a teaching tool that focuses exclusively on the management of the native time cycles: His audience is the European reader. As his written explanation elaborates:

> we have two tables or wheels by which we measure all the [units] of time, one of the golden number, the other of the dominical letter. Thus these *naturales* have two tables, one of 20 figures which serve to count days and months and weeks and years; the other of 52 figures which serves to indicate which year is the current one, and what it is called, and what number it is. (Spitler 2005a, p. 129, after Dyer 1996)

Among the many variations of the Motolinía wheel, one of the earliest appears in the chronicler Muñoz Camargo's *Relaciones Geograficas de Tlaxcala* (hereinafter Codex Tlaxcala) (1585) (Acuña 1984, facing p. 222) (Figure 3.3). Muñoz Camargo's intention seems to have been to liken the 52-year cycle to the Judeo-Christian concept of a specially designated cycle-ending Jubilee year, namely, that terminating the seventh sabbatical cycle of seven years, or the year thereafter (for the

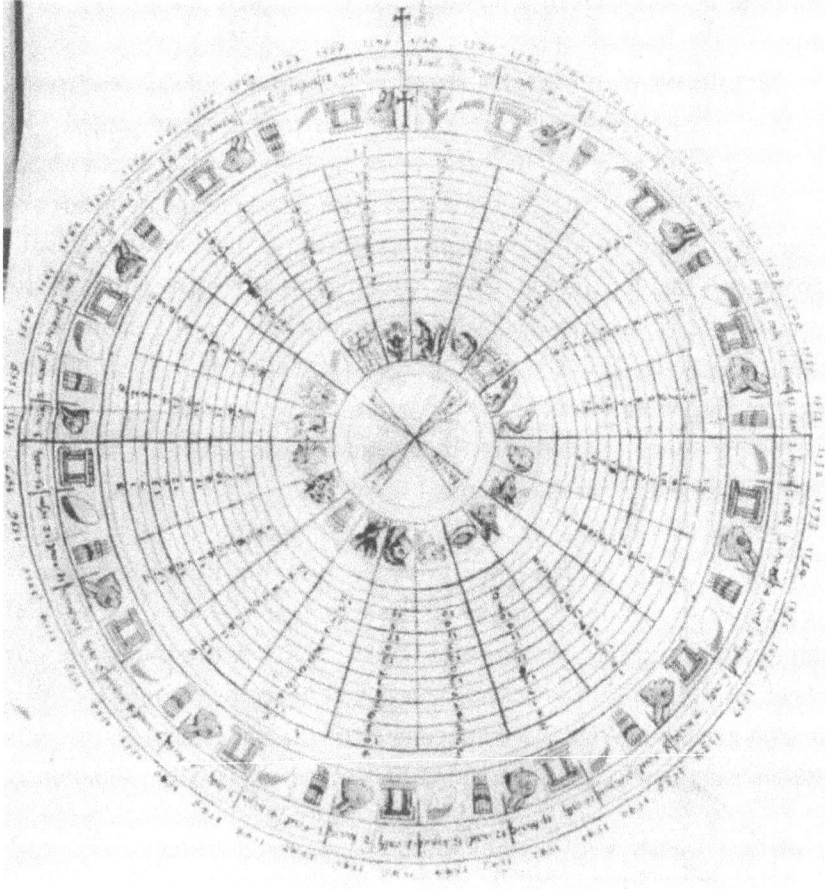

Figure 3.3   Calendar wheel in the Codex Tlaxcala known as Muñoz Camargo Wheel No. 1, (1583–5).

Reprinted from R. Acuña (1984). LR 00521748 (#20398129), with permission of the University of Glasgow.

Christian confession and pardon of sins), the 49- (or 50)-year cycle being a close approximation to a 52-year calendar round.[10] In that part of the text written by the Franciscan friar Francisco de las Navas, he takes credit for having "reduced" the native calendar so as to simplify it for European understanding: "to a wheel of 20 figures, making it a spiral, so that it unrolls [from the center] forward and up" (Baudot 1995, p. 443). In the opinion of D'Olwer (1987, pp. 48–49): "The 1560 calendar composed by Navas is based on the one from 1549, according to the few facts available on it." He cites J.F. Ramirez (1898), who states that Navas, according to the anonymous Tlaxcaltecan who copied it, was the author of a "Calendario indico de los indios del mar oceano y de las partes de este Nuevo mundo, agora nuevamente puesto en forma de rueda para ser mejor entendido . . . ." ["Indian calendar of the Indians of the oceanic sea and the parts of this new world, now newly put into the form of a wheel so as to be better understood."] Ramirez further states that the system "is substantially that of Motolinía."[11]

Interestingly, the spiral aspect evident in the Motolinía Wheel is absent in the Muñoz Camargo rendition, which consists only of concentric circles, though dots along the ring indicate where the user must jump outward to the next ring. Nonetheless, Baudot (1995, p. 485) speculates that the spiral concept might have been alluded to, the better to serve the establishment of the Christianized autonomous native society that lay at the basis of Motolinía's millenarian agenda.

Another difference between the Muñoz Camargo and Motolinía wheels is that all 52 of the outer compartments of the former are filled with names and pictures. Also, Christian years beginning with 1519 (Acatl) and ending with 1570 (Tochtli) are written on the outer periphery of Muñoz Camargo's Tlaxcalan wheel. Finally, the names of the 20 days and the long quote between the 260-day spiral and 52-year compartmented section found in the Motolinía are absent. The Franciscan chronicler Fray Geronimo de Mendieta, who wrote his *Historia Eclesiastica Indiana* in 1596, claims that he had seen a calendar wheel in the convent of Tlaxcala some 40 years earlier. He had arrived in the New World in 1554 and shortly thereafter was installed as bishop of Tlaxcala. Wilkerson (1974, p. 65) speculates that this may have been the wheel that appears in the Motolinía manuscript. A much later version of this wheel appears in Veytia ([1907] 1973, No. 2).[12] It omits the designated Christian years and the central triangles and it reads in the opposite direction.

A highly ornate version of the Motolinía wheel, believed to be contemporaneous with that of Muñoz Camargo, can be found in Fray Diego de Valadés's *Retorica Cristiana* [Lat: Rhetorica Christiana] (1579;

Figure 3.4) (cf. Glass 1975, fig. 73, where it is printed upside down). This wheel is also intended to be read clockwise. Above it, and attached at the position where the Latin cross would have appeared, and where we find instead a compass point, is a smaller wheel consisting of 18 Hispanic-looking faces shown in profile. Each is attached to a tabular arrangement consisting of 9 rows and 4 columns that flanks either side of the wheel. The captions of the four columns read "Nombre de los meses" ["Names of the months"], "Principio del mes" ["Beginning of the month"], "Fin del mes" ["End of the month"], and "El mes del ano de los indios [quantos] dias toma el mes de los cristianos" ["The month of the Indians' year how many days the month of the Christians takes," that is, "the month of the Indians' year correlated with the days of the Christian month"]. In the Rhetorica they are written in Latin. According to Glass (1975, p. 232) the correlation 1 Tlacaxipehualiztli = 1 March is the same as that of Motolinía. Glass notes that a 1579 version of the Valadés wheel published in Italy includes European zodiacal signs. Finally, like that of Muñoz Camargo, Valades's wheel consists of concentric circles rather than spirals.

Among still later variants of the Motolinía wheel is that of Gemelli Careri (Figure 3.5). It is likely from Texcoco (Kubler and Gibson 1951, p. 59). Dated 1697, or more than a century after the wheels in Muñoz Camargo and Valadés, this one is vastly more Europeanized. Its outermost ring is segmented into the four year-bearers, labeled and pictured: Acatl, Tecpatl (*Tecptl*), Calli (*Cagli*, in Italian), and Tochtli (*Tothtl*) and repeated 13 times (in a counterclockwise direction) that sum to 52. The label SECOLO MEXICANO, or "Mexican Century," appears outside and at the top of the diagram. That the calendar is intended to be reckoned counterclockwise is further implied by the order of symbols drawn in the innermost circle beneath the label "Anno Mexicano" in dot/picture notation: 1 Reed, 2 Flint, and 3 House. In place of the anticipated 4 Rabbit appears a curious image resembling a hill with three footprints leading up to it. Interestingly, as the accompanying text explains, these four signs are correlated with the four seasons Tochtli (spring), Acatl (summer), Tecpatl (autumn), and Calli (winter),

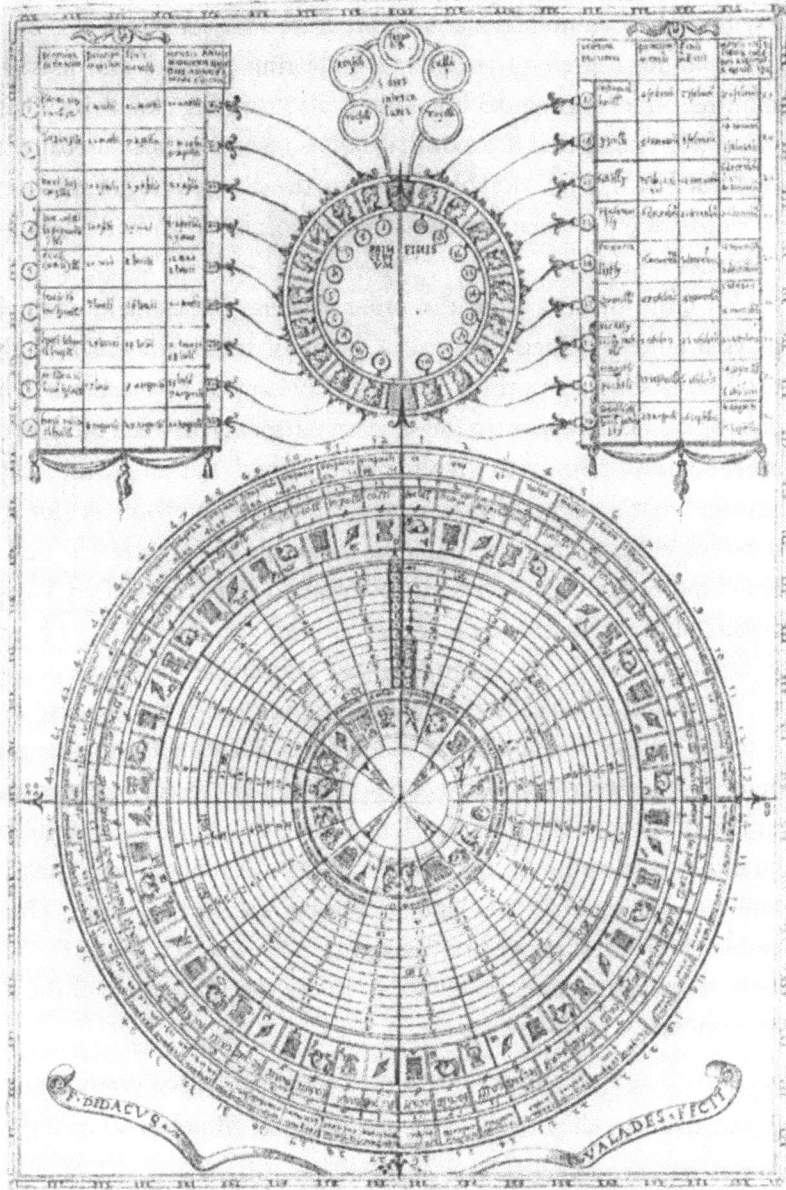

Figure 3.4 Valadés Calendar Wheel (1579).

Reproduced with permission of the Latin American Library, Tulane University.

as well as with the four elements, Earth, Water, Air, and Fire, respectively. The label "Anno Mexicano" ["Mexican Year"] appears at the top
of the innermost circle. The wide, middle ring pictures and lists the
18 months, oddly segmented into sixths, also running clockwise. Glass
(1975, p. 232) notes that the month symbols resemble those that appear
in the Codex Ríos (89r). Notably absent are the detailed rings containing
the *tonalpohualli* days that appeared in the Motolinía, Muñoz Camargo,
and Valadés wheels.

Also quite distinct from the other wheels, the Gemelli example
displays a curious added element: a series of moon drawings in the
form of faces shown in profile and partially capped by crescents. The
crescent-moon faces appear in the six compartments between the central
year-bearers and 18-month glyphs of the *xihuitl*. Although the Maya
did group lunar synodic months in semesters (i.e., units of six) for the
purpose of eclipse prediction (Aveni 2001, p. 157), here the lunar
symbols more likely constitute an attempt to liken lunar Old World
semesters of lunar phases to the indigenous *xihuitl*. On the other hand,
Brotherston (2005, p. 77) speculates that seasonal feasts in relation to
the synodic and sidereal lunar months are intended. Little sense can
be made of the placement of these moon faces. They appear to alternate
more or less between two consecutive individual crescent effigies and
a conflated version of same, the latter possibly intended to indicate
occultation, or new phase. In two instances a doubled image pairs a
frontal sun (or moon?) face and a crescent moon in profile. These
resemble the eclipse images in Sahagún's *Book 7* (1953, fig 21). The
sequence, reading counterclockwise from the top uppermost of the six
compartments, is as follows:

> paired sun and crescent—crescent—crescent—paired crescents
> (interrupted by divider)—crescent—crescent (divider)—
> paired crescents—paired sun and crescent (divider)—
> paired sun (?) and crescent—crescent—crescents
> (interrupted by divider)—crescent—crescent—
> paired crescents (interrupted by divider)—crescent—
> crescent—paired crescents (divider)

Figure 3.5 Gemelli-Careri Wheel, 1697.

Reprinted with permission from J.P. Berthe. (1968). *Le Mexique a la Fin du XVIIe Siecle vu par un Voyageur Italien, Gemelli-Careri*. Paris: Callman-Levy, p. 144 (facing).

Counted in this manner there are 17 entries, though the total number of images appears to be 23. Some sort of correlation between lunar months of 29/30 days and the indigenous 18 x 20 + 5-day months in the indigenous calendar may have been intended. The closest commensuration between the Mesoamerican 20-day month and the lunar synodic period is the two-thirds ratio: 6 x 30 = 9 x 20 = 180, which rounds off the lunar half year with a phase difference of approximately minus three days per semester. If this is the equation being sought, it is hardly functional in the form rendered in the Gemelli wheel.

A curious statement about the moon in relation to Venus in the accompanying text reads: "Distinguían el movimiento de la luna en dos tiempos: el primero de vigilia, desde el Orto Heliaco o nacimiento del sol, hasta la oposicion, de (catorce) dias, y el Segundo de sueño, de otros tantos dias hasta la ocultacion matutina". ["They separated the movement of the moon into two periods: the first in the waking hours, from the heliacal rising or the appearance of the sun, to the opposition, of fourteen days, and the second in the sleeping hours, of the same number of days until the morning disappearance."] These two periods are roughly equivalent to what we call the waxing and waning phases of the moon. The text goes on to suggest a connection between the number 13 and lunar-phase reckoning: "después de trece dias de haber llegado el sol al occidente, se iba descubriendo la luna en oriente" (Perujo 1976, p. 48)[13] ["thirteen days after the sun had reached the west, the moon started to appear in the east"].[14]

Finally, the Gemelli wheel is encircled by a coiled snake arranged head to tail in the clockwise direction, with loops positioned at each of the four directions, and the head meeting the tail at the top of the picture. This form is most likely (though not necessarily) derived from the Greek *ouroboros*, a symbol of eternity, or that which restarts itself at the point of completion. A drawing of a serpent devouring its own tail appears in the reportorio of Chavez (1581 f.27v) (Figure 3.6), which may have served as the prototype.[15] A wheel much like the Gemelli, with a crescent moon at the center, appears in Veytia ([1907] 1973, No. 4), however, it does not include the phases. Additionally, some of the month names and symbols appear to be out of order. In another version found in the Boturini collection (Oudijk and Castañeda de la Paz 2010, fig. 7, #8), a sun face occupies the center.

A second wheel in the Codex Tlaxcala, facing p. 223, appears to be the prototype for another series of Motolinía copies (Figure 3.7). It

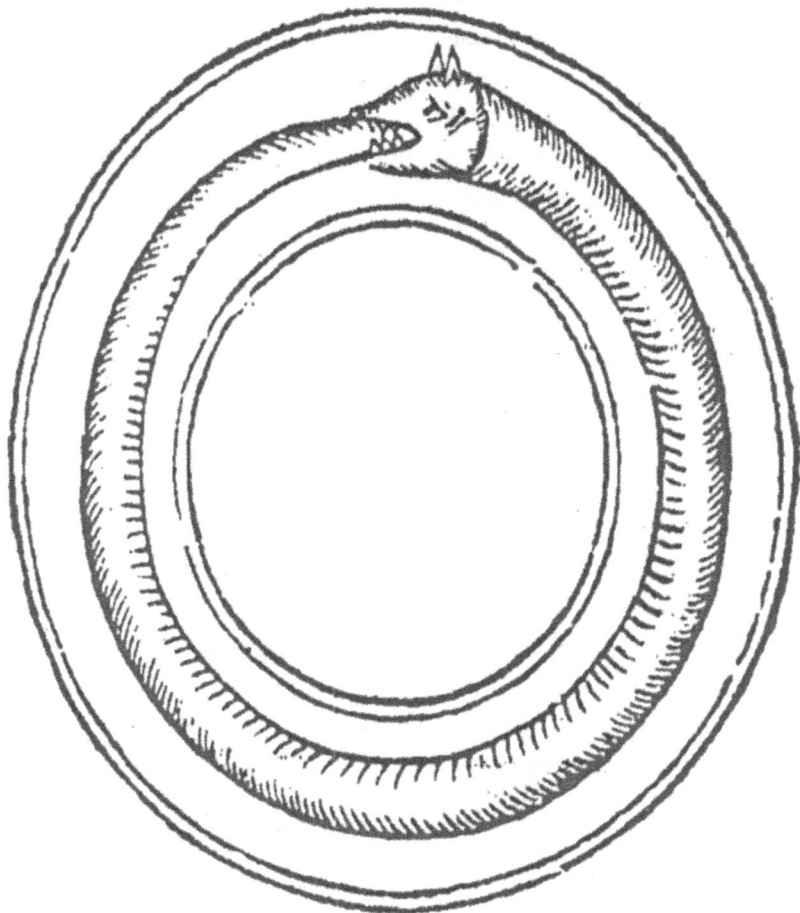

Figure 3.6 Representation of a coiled snake swallowing its tail from a Spanish Reportorio.

Reprinted from J. de Chavez. (1581). *Cronographía o Reportorio de los Tiempos, El Más Copioso y Preciso que Hasta Ahora Ha Salido a Luz.* Seville: Juan Francisco de Cisneros, p. 27.

depicts Europeanized (though not as obvious as the seventeenth-century Gemelli) symbols of the 18 months, with the name and number of days in each written along the periphery. An added compartment consisting of five dots in Aztec codex style is labeled "Dias intercalares" ["intercalary days"]. Brief texts below each symbol indicate the nature of the feast celebrated in that particular month. References to the *tonalpohualli* are absent. The inner ring consists of a large crescent-moon face shown in three-quarters frontal, possibly eclipsing the sun, though Milbrath (personal communication, 2/7/11) points out that this clearly does not represent the stylistic convention one sees in eclipse representations, for example, in Sahagún, *Book 7*. An inscription appearing in a ring circumscribed about the moon reads: "Quenta y orden de los meses de todo el año atribuidos a la luna que un mes de veynte dias llaman una luna" ["count and order of the months of the whole year attributed to the moon for they called a month of twenty days one moon"]. Veytia No. 5 (Figure 3.8) looks like a copy of Muñoz Camargo's Wheel No. 2, though Glass (1975, pp. 233–4) believes the latter to be a copy of a wheel now lost from the Boturini collection.

In the Lorenzana version of Muñoz Camargo No. 2 (1770) a crescent moon is situated at the center and the five-dot segment is smaller, thus more accurately representing the actual length of the *nemontemi* when placed alongside the 20-day months (Figure 3.9; Glass 1975, fig. 75). Also, the interpretive statements attached to each of the months are briefer. The legend at the bottom states that there are 18 months of 20 days in the Mexican year, which total 360 days, that 5 days are added to complete the 365-day year, and that this is the calendar they use to compute their chronology. There follows the phrase ". . . no para las Obserbaciones Astronomicas" [. . . not for Astronomical Observations"]. The inscription that envelops the crescent-moon face is identical to that in Muñoz Camargo No. 2. A *croix formée*, which frequently serves as a completion symbol in Central Mexican calendars (Aveni 2000), is positioned beneath the *nemontemi* compartment at the start of the sentence in Veytia 5, but not in the other variants, to indicate where the count begins. Glass (1975, p. 233) notes that the beginning point, Atemoztli, is a Tlaxcalan trait. Another version, which resembles Figures 3.7 and 3.8, is AGN 198, from the Historia (Vol. 1, Exp. 19); it dates from 1790 and has the months positioned in the usual manner. The insistence of placing the figure of the moon at the center of each of the wheels of this group further perpetuates the Hispanic implication

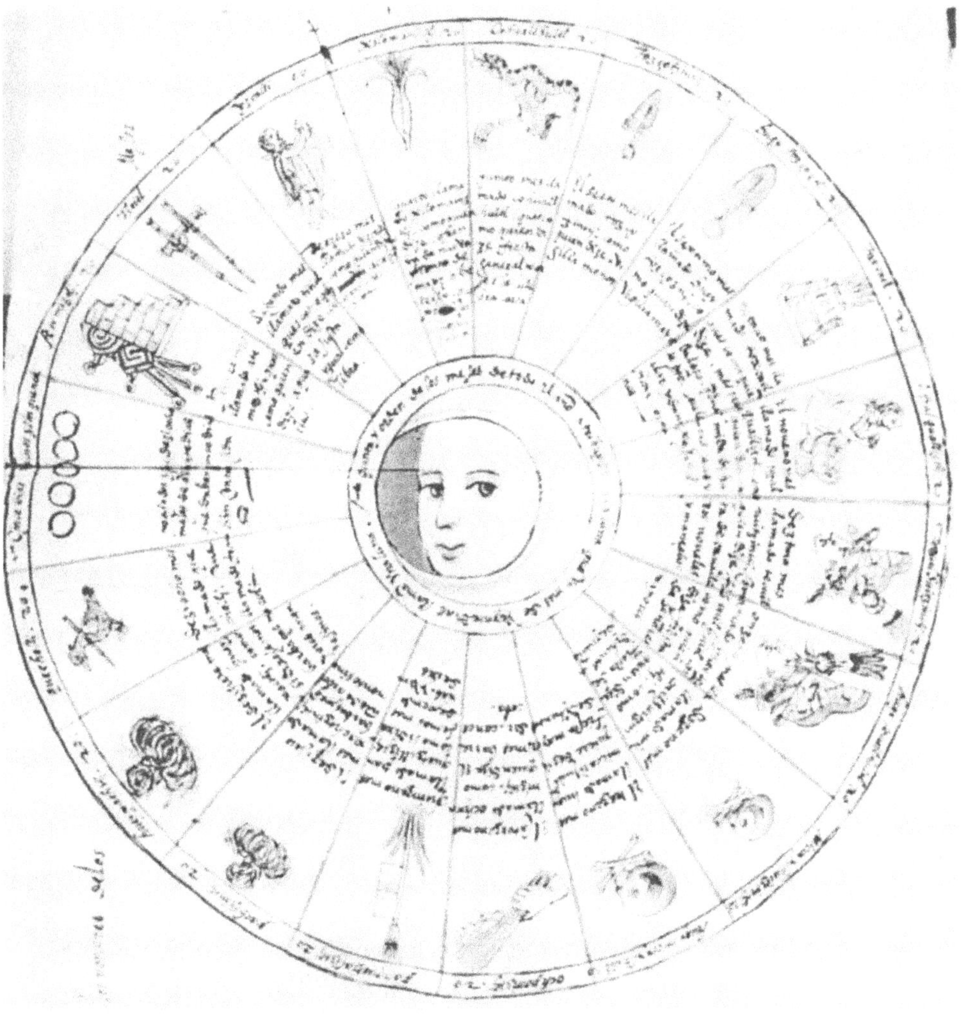

Figure 3.7. Muñoz Camargo Wheel No 2. Note the crease showing where it had been folded.

Reprinted from R. Acuña (1984). LR00521748 (#20398128 H242), with permission of the University of Glasgow.

that the Mesoamerican months must be somehow connected with the lunar synodic period. For example, another (undated) wheel from the Boturini collection (Oudijk and Castañeda de la Paz 2010, fig. 6, #5) displays the crescent moon at the center. As in the Gemelli wheel, the

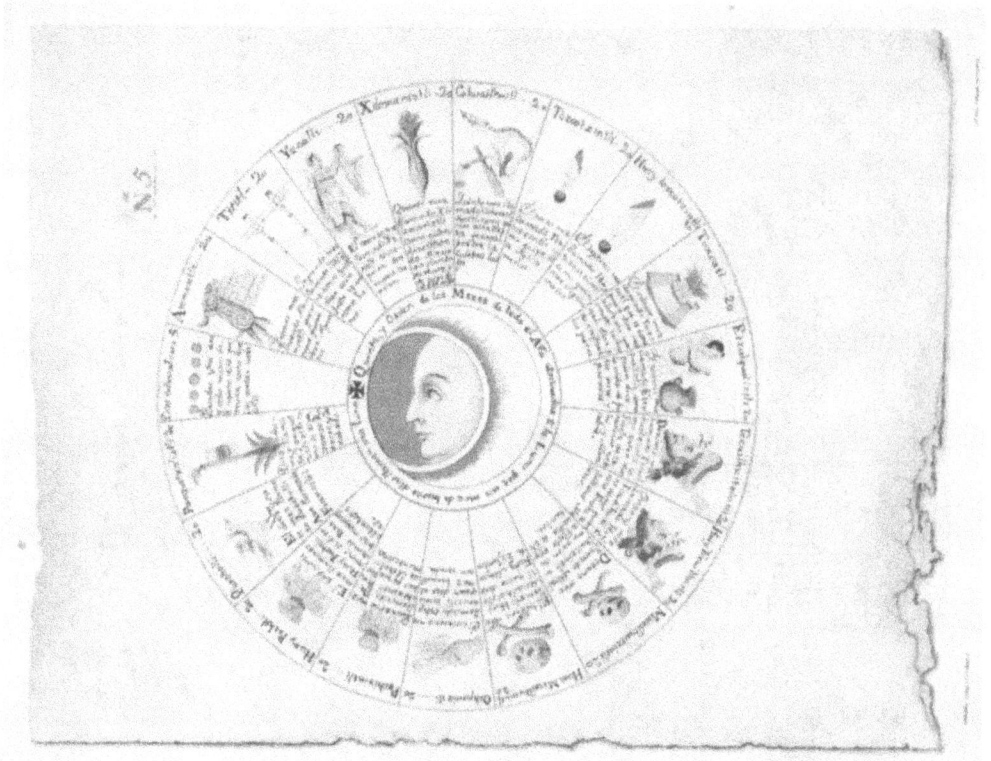

Figure 3.8. Veytia Wheel No. 5.

Reprinted from M. Veytia. [1907] (1973). *Los Calendarios Mexicanos* Mexico: Museo Nacional.

outer compartments are divided in sixths with three months sharing each compartment; the *nemontemi* are omitted.[16] A second Boturini example (Oudijk and Castañeda de la Paz, fig. 6, #6), labeled "Rueda de los Dias del Año Civil" ["Wheel of the Days of the Civil Year"], emphasizes the quadripartite structure via four spokes that emanate from a central hub.

Another wheel with the so-called "ouroboros" structure, though otherwise quite unlike the Gemelli Careri, resides in the collection of the Peabody Museum, Harvard (PMAE Number 971-1-20/23936) (Figure 3.10), where it is incorrectly labeled Vaytia [*sic*] calendar wheel #1. Not mentioned in the Glass survey, this wheel consists of 19 wedge-

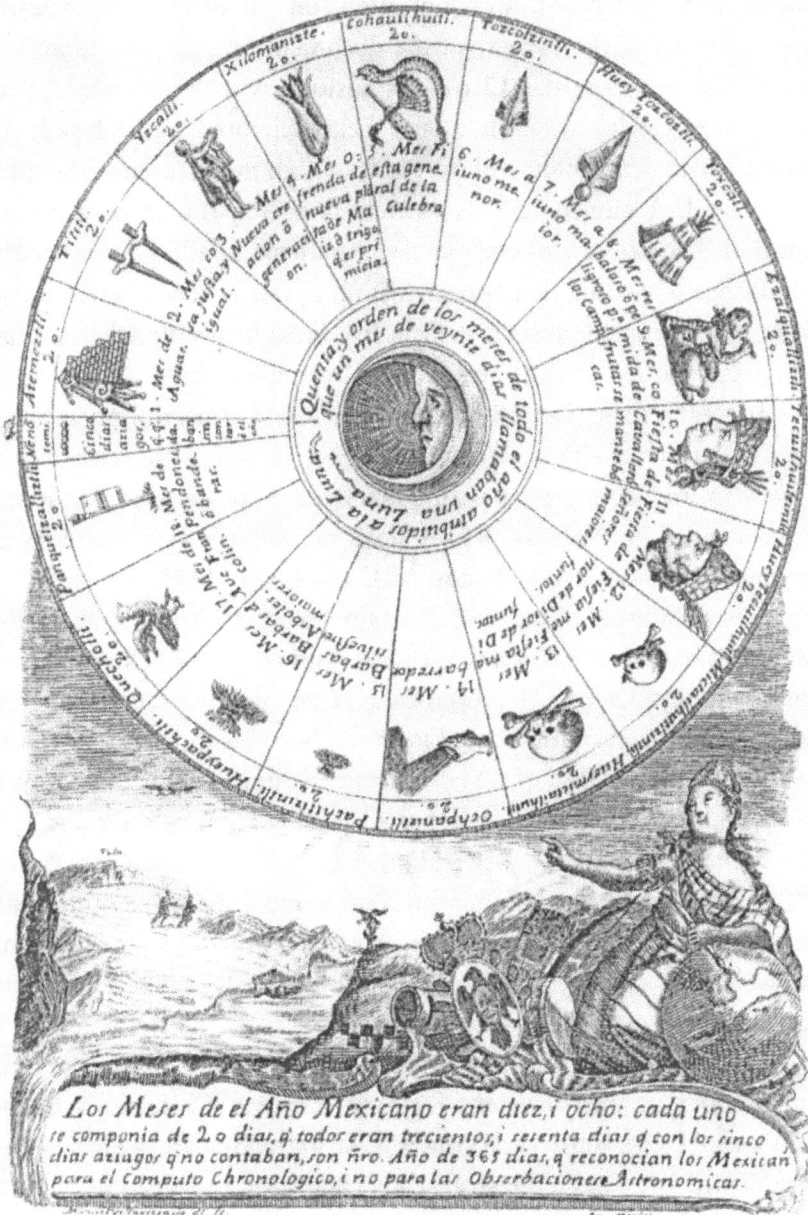

Figure 3.9 Lorenzana Wheel (1770).

shaped compartments that radiate outward from the center, which houses a Tecpatl year-bearer indication, one of the four year-bearer signs pictured at the corners of the document outside the circle. The month list begins at the 12 o'clock position with "Atemochtli" [*sic*] and moves clockwise, each month being pictured and labeled. An accompanying explanation of the feasts of each month appears beneath the symbol. A circled "20" is written along the rim at each month position. The *nemontemi* compartment, perhaps to allow room for the abundance of writing that appears within it, is accorded a space somewhat larger than the rest; it depicts five faces in profile labeled "dias complementarios" ["additional days"]. Decorative panels jut out from the 3rd, 8th, 13th, and 18th compartments. An inscription in large cursive script on either side of the wheel identifies the calendar as Tlaxcalan and connects it with Franciscan Fray Joan de Benavente (Motolinía). The Peabody Wheel looks more like a very late copy of Veytia's Wheel No. 5 (see Figure 3.8), and is very likely a forgery.[17]

Veytia Wheels No. 6 and 7 retain mixed aspects of the earlier Motolinía and Muñoz Camargo wheels, with added embellishments of their own. Veytia Wheel No. 7 (Figure 3.11; **see also color plate opposite p. 56**), from Tlaxcala, and dated to the early eighteenth century (Glass 1975, p. 234), consists of 20 compartments numbered on the outside 1–13, followed by 1–7, with repeated pictured symbols for the four alternating year-bearers, labeled bilingually "Ce Calli (Una Cassa)" [*sic*] ["One House"], and so on. Beneath each symbol the three other year-bearer symbols follow in miniature; thus, under the drawing of a rabbit in the *Tochtli* compartment appear: Acatl, Tecpatl, and Calli, and so forth. Written over each of the symbols, respectively, are the letters V, E, O, H, which, according to the fully written-out versions in the Calli compartment, stand for Verano ["Spring"], Estival ["Summer"], Otono ["Autumn"], and Hibierno ["Winter"]. As in the Gemelli–Careri Wheel (see Figure 3.5, p. 29), the year-bearers appear to be forced into representing the four European seasons of the year, even though no temporal correspondence among these intervals is indicated. Nor is there any known indigenous conception of the year having been divided into

Figure 3.10 Possible reproduction of the Veytia calendar wheel #5. Manuscript on parchment.

four seasons, either astronomically or meteorologically.[18] Twenty inner compartments, also numbered 1–13 followed by 1–7, with the numbering system backshifted by two places with respect to the year-bearer compartments, name and picture the 20 days of the *tonalpohualli*. The order commences with 1 Cipactli (incorrectly labeled Tiburón, or "Shark," instead of crocodile). The inner circle pictures a radiating frontal sun face at the left, a profile crescent-moon face at the right, a single star at the top, and four stars in the shape of a cross at the bottom. The intent may have been to correlate the diagram with the four directions. If one were to conceive of the inner diagram as a sky map in the European sense (holding the map overhead), the sun would be positioned in the east, the moon in the west, the North Star in the north, and the four stars (possibly the constellation of the Southern Cross) in a southerly direction. For a discussion of the variants of the Veytia wheels in the editions of 1836, 1907, and 1994, see Glass (1975, pp. 229–34).

Veytia No. 6 (color illustration) is a rendition of the inner portions of Veytia No. 7. It contains the same celestial symbols as No. 7. These appear to be labeled A B C D just outside the outermost ring next to one each of the year-bearer signs. That ring counts the coefficients 1–13 and 1–7 in Aztec dot rather than Arabic notation, the count beginning on 1 Cipactli (this time more correctly drawn, though incorrectly labeled "serpiente"), which is positioned at the ca. 2 o'clock rather than at the ca. 11 o'clock position, as was the case in Veytia No. 7. A frontal sun face replaces the more archaic symbol in the (4) Ollin compartment, the date that was to end the epoch of the "Fifth Sun" according to Aztec lore.

The Veytia wheel No. 1 dates from 1654 and more closely resembles the Motolinía wheel (Figure 3.12). It correlates the 52-year cycle with the years 1649–1700 (Glass 1975, p. 230). Four year-bearer signs appear in quarters divided by lines at the center. Outer compartments commence at the 9 o'clock position, with the house symbol labeled "1649" on the outer ring, then "1650" (the rabbit symbol), and so on. The year-bearers are labeled 1–13 and repeat four times on the next inner ring, for a total of 52 compartments. A label positioned over the year 1654 (6 Rabbit) reads "now, in this year 1654" (Glass 1975, pp. 230–1). A rectangular table positioned above the wheel radiates via several lines from the inscribed year 1663. It indicates New Fire ceremonies celebrated in the 2 Reed years dating back to 1195, 1663 being one such year. All but the first four entries in the table are left blank. Entries consist of place names and symbols, for example, Tepayocan, Chapoltepec [Chapultepec] . . . . These were stop-off places on the Aztec migration. The number two (a pair of Aztec dots) for the second Aztec New Fire Ceremony appears over the 1195 year and place entry, which is labeled A; the other three carry labels B, C, and D.

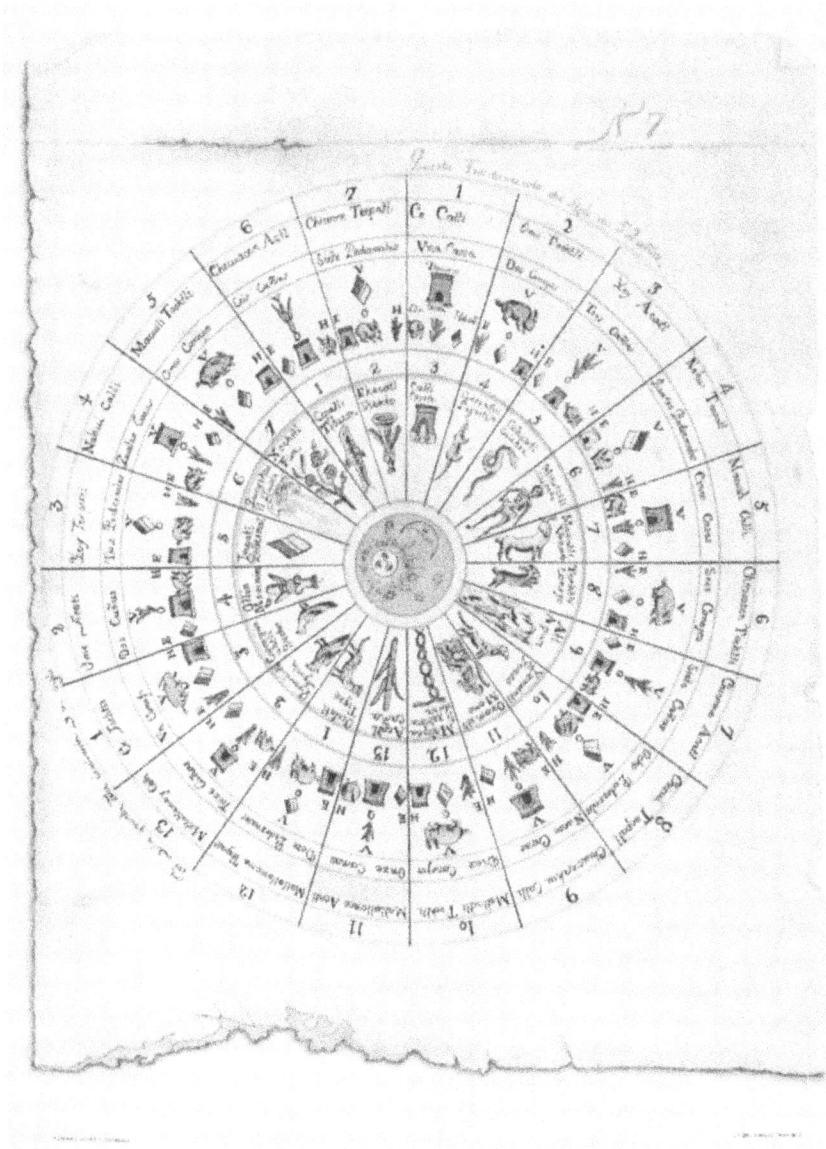

Figure 3.11 Veytia Wheel No. 7.

Reprinted from M. Veytia [1907] (1973). *Los Calendarios Mexicanos* Mexico: Museo Nacional.

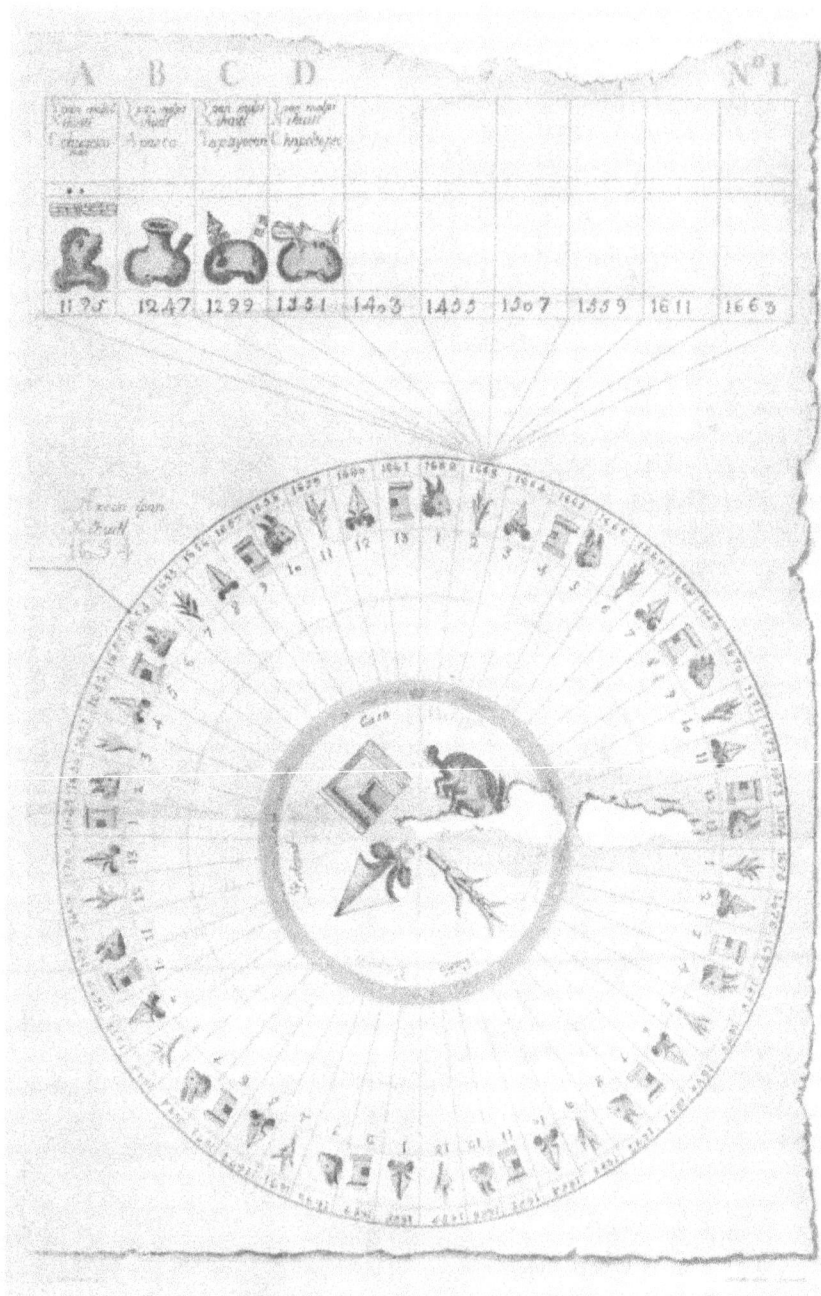

Figure 3.12   Veytia Wheel No. 1.

Reprinted from M. Veytia [1907] (1973). *Los Calendarios Mexicanos* Mexico: Museo Nacional.

Because the year 1654 coincided with the writing of Jacinto de la Serna's *Manual de Ministros de Indios* ([1654] 1892), it is likely that this wheel relates to a version found therein. Serna's wheel No. 2[19] (Figure 3.13), however, runs counterclockwise and, instead of listing year-bearers, it pictures and labels only the days and numbers connected with the 260-day calendar, running from 1 Cipactli through 7 Xochitl. Thus it constitutes a radical departure from the other Mexican wheels, resembling more closely those from postconquest Maya texts. The Serna wheel is further squared off by a boundary, the four corners of which carry the year-bearer symbols with labels "Meridies" (a house pictured at the upper left), "Septe[n]trio" (rabbit to the lower left), "Occide[n]s" (reed), lower right, and "Oriens" (flint) upper right, a directional association that conflicts with other diagrams that will be described in the next section, where Reed is to the east, Flint to the north, and so on.

The Calendar Wheels of Sahagún

The second earliest calendar wheel from Central Mexico appears in *Book 7* of Sahagún's *Florentine Codex*, "The Sun, Moon, Stars, and the Binding of the Years" (Figure 3.14). It was copied from the slightly more detailed version found in his Tlatelolco Manuscript published between 1563 and 1565 (cf. Glass 1975, p.189), and was likely painted by native multilingual artists educated in the European tradition. The wheel displays the 52-year cycle via the four year-bearer dates of the *tonalpohualli*. Sahagún (1953, *Book 7*, opp. fig. 20) calls the wheel a *table* and he describes it as follows:

> This table ... is the year count, and it is a most ancient thing. They say that its inventor was Quetzalcoatl. It proceedeth in this way: they begin with the east, which is where the reeds are (or, according to others, with the south, where the rabbit is) and say One Reed. And thence they go to the north, where the flint is, and they say Two Flint Knife. Then they go to the west, where the house is, and there they say Three House. And then they go to the [south], which is where the rabbit is, and they say Four Rabbit. And then they turn to the east, and say Five Reed. And thus they

Figure 3.13   Serna Wheel No. 2.

Reprinted from J. de la Serna. [1654] (1892)."Manual de Ministros de Indíos para el Conocimiento de sus Idolatrias, y Extirpación de Ellas." In *Tratado de las Supersticiones, Dioses, Ritos, Hechicerias, y Otras Costumbres Gentilicas de las Razas Aborigenes de Mexico*, p. 146. Mexico: Ed. Fuente Cultural.

libro 7° de la astrologia

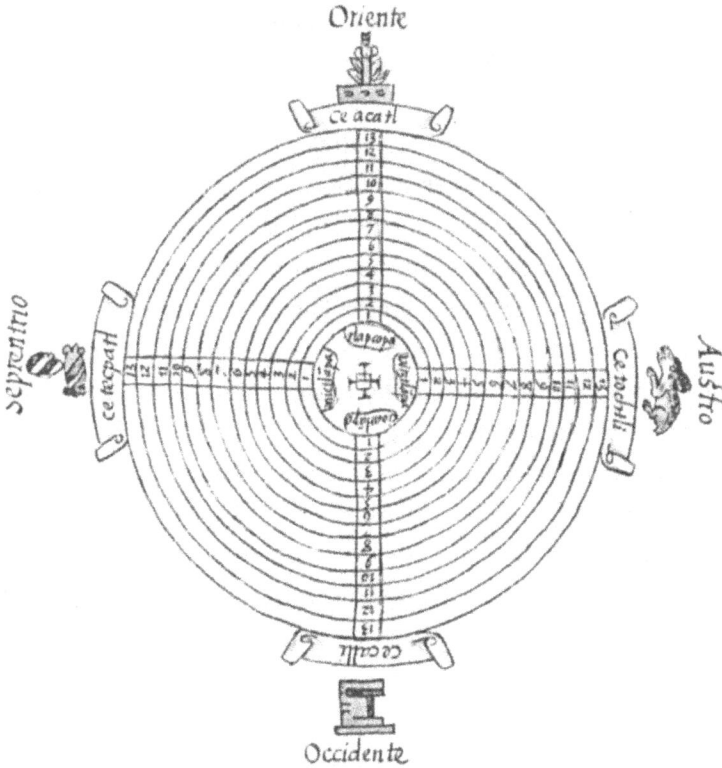

Oriente

Septentrio

Austro

Occidente

Figure 3.14   Sahagún Calendar Wheel.

Reprinted from B. de Sahagún. (1953). *Florentine Codex* Book 7, fig. 20, with permission of the University of Utah Press.

go, making four revolutions, until they reach thirteen, so that they end where they began. And then they return to one, saying One Flint Knife. And in this way, making revolutions, they assign thirteen years to each of the characters, or to each of the four quarters of the world. And then fifty-two years are completed, which is a bundle of years, when the Jubilee is celebrated

and new fire is made in the manner set forth above. Then they again count
as at the beginning.

Unlike Motolinía's wheel, which takes on the shape of a spiral,
Sahagún's wheel consists of 13 concentric shells. Pictures of the four
year-bearers: Rabbit, Reed, Flint, and House (also written in Nahuatl)
are assigned to the cardinal directions: south, east, north, and west,
respectively (written in Spanish); east is positioned at the top. Despite
the circular pattern, time flows in *spiral* fashion in a counterclockwise
direction, beginning with 1 Rabbit in the innermost circle, then proceeding to 2 Reed in the second circle, 3 Flint in the third, 4 House, 5
Rabbit, and so on. A complete cycle of 52 years places the user in year
13 House on the outermost circle. The four quarter directions are
inscribed thus: *uitztla[n]pa* (south), *tlapcopa* (east), *mictla[n]pa* (north),
and *cioatla[n]pa* (west), inside the center circle surrounds a figure of
a cross centered on a square.

The Relacíon de Meztítlan, written in 1579 (Acuña 1986), states
that Gabriel Chavez, the Mayor Alcalde [Primary Chief] of Meztítlan,
and a Creole, noted that they began the count with One Rabbit in the
east, and then proceeded to the north and west, "andando [a] la *redonda*
hasta, trece años" (p. 56, my italics) ["moving in a circular fashion
until (it completes) thirteen years"]. Here the word "redonda" replaces
"vuelta." Later (p. 75) the Relacíon says that "La cual di[ch]a relacíon
yo, el d[ic]ho GABRIEL DE CHAVEZ, hice con informacíon muy particular de los indios mas viejos q[ue] hallé en la d[ic]ha provi[nc]ia, de
los cuales, y de personas antiguas, me informé de todo lo contienda en
esta relacíon; y me constó por PINTURAS ANTIGUAS q[ue] yo vi . . . "
["This said *relación* I, the said GABRIEL DE CHAVEZ, prepared with
very particular information from the oldest Indians I found in said
province, from whom, and from very old persons, I informed myself
of all the contents of this relación; and it was recorded in ANCIENT
PAINTINGS which I saw . . . "] (E. Wake [personal communication,
11/25/10] notes that "dando bueltas," the operative words, can also
mean to go around something not necessarily circular, that is, in the
sense of tracing a complete path where the start is the same as the

finish. Thus Sahagún may well have been working with a table, as he so categorizes his wheel, but his terminology likely prompted whoever was making the pictorial to convert it into a European wheel. Therefore, the Spanish text is ambiguous as far as the shape of the artifact is concerned.)

Miram and Bricker (1996, p. 402) and Spitler (2005b, p. 276, and fig. 16-3) liken the Sahagúntine wheel to European diagrams that show the Earth-centered universe, with planetary orbs arranged in different layers of heaven inside a firmament consisting of star symbols arranged according to the signs of the zodiac, with a drawing of the Earth at the center. Additionally, Robertson (1959, 1966), in exploring the relationship between Sahagún and his informants, has compared the *Florentine Codex* with the *De Proprietatibus* of Bartholomaeus Anglicus, suggesting that there may have been a copy of this widely circulated book in the library at Santa Cruz de Tlatelolco, where Sahagún did much of his work. My own search of the inventories failed to turn up this particular reference, though the reportorio of Chavez is listed in Garcia Icazbalceta's (1892) census, which lists inventories for 1572, 1574, and 1584 (cf. also Mathes 1982).

Sahagún's wheel bears a distinct resemblance to the Copernican diagram of the solar system (see Figure 2.1a, p. 12). Like the Sahagúntine wheel, the Copernican diagram emphasizes the concentric circular zones in which the planets move. It is worth noting that Sahagún's career (1499–1590), like most of the early chroniclers, overlapped with that of Copernicus (1473–1543). Educated at the University of Salamanca, Sahagún arrived in Veracruz in 1529, well before the publication of *De Revolutionibus*. It is difficult to know what degree of personal access he may have acquired to the then avant garde European theories of cosmology (his *wheel* gives no hint of a sun at the center, which instead houses a cross), nevertheless he and his native assistants certainly made use of similar diagrams in the reportorios. Lopez Austin (1974, p. 135) makes a strong case that *Book 7*, in which the wheel appears, is especially filled with misconceptions about the subject of Aztec natural philosophy, Sahagún having been wedded to totally occidental expectations of answers to questions he posed to his informants.

Interestingly, in *Book 5* Sahagún (1979, pp. 139–40) quotes from a text by Motolinía that he then proceeds to criticize. This text is worth reproducing in its entirety because of the insight it offers us on the matter of how these early chroniclers quibbled over the conceptualization of the native calendar:

> By means of the wheels reproduced above, the Indians count their days, weeks, months, years, Olympiads, five-year periods, indictions, and hebdomads,[20] starting their year, like ours, with the beginning of January. In [their system] are found the ways of counting the times which all nations have had. And it seems the Indians who established and knew it well certainly showed themselves to be natural philosophers. They were at fault only as to the leap year. But this was the case also as to the great philosopher Aristotle and his teacher Plato, and many other wise men who did not attain [knowledge of] it. And it should be known that in this calendar there is nothing idolatrous; and this is praiseworthy for many reasons, but to mention one will be enough. And it is that in this land it was not very many years ago that idolatry began. And this calendar is very ancient, and if the names of the days, weeks, and years, and their representations, are of animals and beasts, and other creatures, it is nothing to wonder at; for if we look at ours, they also are of planets and of gods which the pagans had. Even if many rites, falsehoods, and ancient sacrifices are written of her [in connection with] a thing so good, of such excellence and truth as these natives possessed, there is no good reason to condemn it. For we know that all good and truth, no matter who expresseth it, is of the Holy Ghost.

Then follows Sahagún's refutation of Motolinía:

> As to what he first saith, that by means of this count the Indians reckoned [their] weeks, months, and years: this is most false. For this count containeth only two hundred and sixty days, and lacketh one hundred and five days to be the count of an entire year. Neither did they reckon their months by this count, because their months are eighteen in a year, and each one hath twenty days, which maketh three hundred and sixty days. This count doth not reach that number. Neither do they reckon their weeks by this count; for what they say, that they had thirteen days in a week, is false, because thus there would be a week of thirteen days and another week [which] would go into the next month with three days, and so no month would have two entire weeks. But more important, their weeks were of five days, which were better called five-day periods than weeks, and there are in each month four of these five-day periods.

What he saith of Olympiads, five-year periods, indictions, and the like, is false and pure invention.

What he saith, that the year started in January, like ours, is most incorrect; for what they call a year by this count is of only two hundred and sixty days, and of necessity it would have to end one hundred and five days before our year. And so it could not start with our year, except with some, and very rarely.

As to what he saith, that the Indians [who] devised this count showed themselves to be natural philosophers: this is most false. For they do not carry out this count according to any natural order; for it was an invention of the devil and an art of soothsaying.

As to what he saith, that they lacked the leap year, it is wrong; because in the count which may be called a true calendar they count three hundred and sixty-five days, and once every four years they counted three hundred and sixty-six days with a feast which for this reason they observed every four years.

As to what he saith, that in this calendar there is no idolatry, it is a very great lie. For it is no calendar but a soothsaying device in which are contained a great deal of idolatry, many superstitions, and many invocations to the demons, tacitly and openly, as is shown in all of this preceding Fourth Book. So that the treatise aforementioned, which that member of a religious Order wrote, containeth no truth but rather very pernicious error and false-hood. (pp. 140–41)

Among Sahagún's acrid criticisms of Motolinía appears the false state-ment that the Aztecs employed the leap year. Additionally, he seems to confuse the *tonalpohualli* and the *xihuitl*.

Codex Mexicanus Calendar Wheels

As alluded to previously, one of the most obvious examples of mechani-cal-looking Mesoamerican time wheels that mesh together appears on page 9 of the Codex Mexicanus (see Figure 1.3, p. 6). It is also one of the earliest examples, dated 1579, or just three years before the institution of the Gregorian Calendar Reform in Europe (see Brotherston 2005, pp. 81–85 for a detailed discussion). Used to indicate moveable feast days, this two-wheel instrument with the figure of St. Peter holding his keys at the center, seems to have been intended to function as a means of locating successive New Year's Days in Old and New World calendars. At a conceptual level the idea of matching day names in the *tonalpohualli* and *xiuhmolpilli* in the Aztec calendar is likened to the European cycle that matches the name of the weekday with numbered day of the month in the Christian calendar.

The principal ecclesiastical calendric problem affecting calendar reform in Europe at the time had to do with connecting the week to the year, specifically to find the day of the week that corresponded to a given date in the year. This was a major issue regarding the location of the Paschal celebration in the seasonal year. If a common year begins on Sunday, for example, the next year will begin on Monday. Should that first year be a leap year, then the following year would begin on Tuesday. For convenience, the days of the week were denoted by the first seven letters of the alphabet, A–G; thus A would stand opposite the first day of January 1, B opposite January 2, etc., through 365 days of the year. If one of the days of the week, say Sunday, were represented by E, Monday would be represented by F, Tuesday by G, Wednesday by A, and so on. Every Sunday of that year would correspond to an E day, the so-called Dominical (Sunday) Letter for that year. Once the dominical letter of a given year was set, the letters of all the other days of the week would become known. Because the number of years between intercalary leap-days is four and the days of the week number seven, the product of the two, 28 years, gave the length of the cycle that covers all possible combinations of the days of the week with the commencement of the year. This is why the red-colored wheel on the left side of Mexicanus 9 contains 28 compartments (not including the cross at the top), thus depicting the 28-year solar cycle, each bearing a letter ranging from A to G, read in a counterclockwise direction. Recessed letters appear at every fourth position, reminding users to skip a letter for leap year. This is because the name of successive New Year days advances by an extra day following a leap year.

The wheel on the right side of Mexicanus 9 represents the Mexican year-bearers. These are arranged in the usual 4 x 13 = 52 *xiuhmolpilli* pattern in a circular format (Rabbit–Reed–Flint–House). That the wheels are tangent to one another might indicate an intention to correlate the letter representing the start of a given Christian year with that in the *xihuitl*; thus the letter C on the red wheel touches 1 Rabbit on the blue wheel. Now, we know that 1576, three years before the date of the manuscript, was a C year (= Friday, 1 Jan.); however, the most recent 1 Rabbit year would have been 1558. Clearly there is considerable slippage going on. Notwithstanding, it is hardly likely that the paired wheels are functional in the form in which they are represented; for example, if the left wheel turns clockwise, the right wheel would be required to turn counterclockwise in order to engage it. If this were the case, the native time count would run in reverse order: Rabbit–House–Flint–Reed.

To the left of the contiguous wheels appears a 365-day feast count (18 x 20 + 5) and (at the top left) the *nahuatl* gloss *naupoualixihuitl*, or "four-year count," possibly denoting the four-year structure shared by the Mexican and Christian calendars; that is, the four-year-bearer character of the former and the four-year leap year structure of the latter (Brotherston 2005, p. 82). A cryptic table follows on page 10 of Codex Mexicanus. It depicts a series of 27 alphabetic entries arranged in a 28 vertical (year) by 18 horizontal (Mexica month) grid, with the twelve zodiacal signs parsed out in two and three compartmental spaces along the vertical edge.

The Durán and Tovar Calendar Wheels

The Durán Calendar Wheel (Horcasitas and Heyden 1971, fig. 35), dated ca. 1581, displays a 52-year cycle named by the year-bearers and arranged in four arms of a cross, each arm bent into a circle at the periphery to form a swastika (Figure 3.15; **see also color plate opposite p. 56**). The arms are labeled cardinally in Spanish. Four cherub-like faces blow winds from the intercardinal directions (cf. Fig, 2.1c, p. 15). Unlike all the wheels discussed previously, a sun face illuminates the center. Cañas (Reed) years open to the east, which is positioned at the top, pedernales (Flint) to the north, casas (House) to the west, and conejos (Rabbit) to the south. The heliocentric symbol lends a spatial element to the diagram. It also may have served the purpose of stressing the primacy of the solar year. The accompanying narrative, as that in Motolinía's text, justifies the use of the circular form; thus:

> The curious reader interested in discovering what this circular design means will understand and easily comprehend what the characters and symbols signify. It simply teaches us to understand the way in which the years were counted by the natives in older times. (p. 388)

Beginning in the east with the innermost day name 1 Caña, one proceeds counterclockwise to 2 Pedernal, 3 Casa, and 4 Conejo, then spirals outward to successive day names along each axis, as in the Motolinía wheel, to 5 Caña, 6 Pedernal, and so on, finally terminating at the tip of the southern axis of the swastika with 13 Conejo. The Tovar Wheel (cf. Kubler and Gibson 1951) is a faithful copy of the Durán and exhibits few significant differences either in appearance or operation.

The Boban Calendar Wheel

The Boban, a detailed, ornate wheel from Texcoco, appears to have been created by a native artist and designed for native use (Figure 3.16). Its form is quadripartite, the ring space on the periphery of the circle containing the 18 month symbols with names written in *nahuatl*. This arrangement mirrors the one displayed in European calendar wheels (cf. e.g., Figure 2.1b, p. 14) with the Mesoamerican months replacing the signs of the zodiac (cf. also Spitler 2005a, p. 161). A cumulative count of days in the *xihuitl*, beginning at the 1 o'clock position and moving clockwise, is printed on the outside of the month ring. Footprints around the inner rim of the calendar indicate the direction of time's movement. They enter at the 1 o'clock position and exit at the 11 o'clock position. Footprints, ubiquitous in the Central Mexican codices, are frequently employed to represent movement in space but

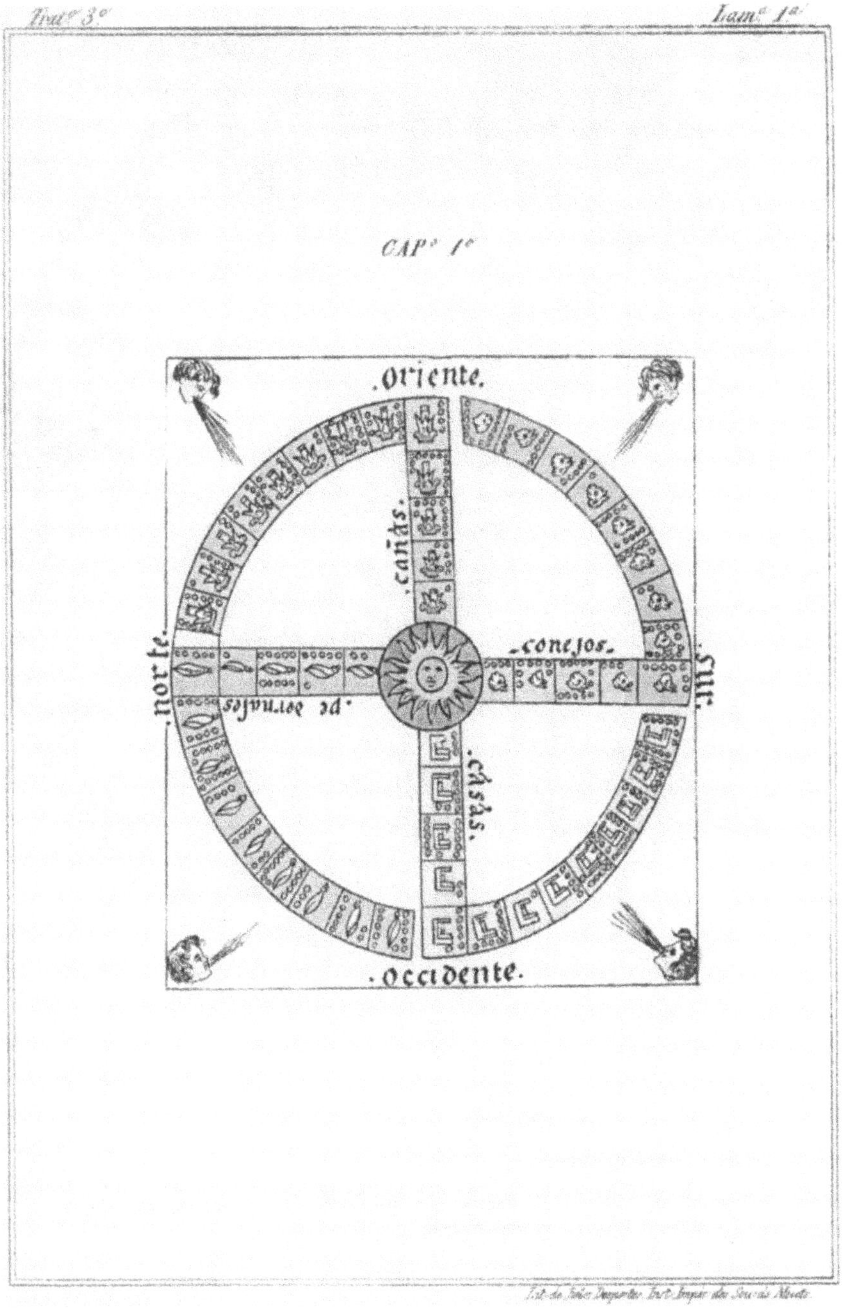

Figure 3.15 Durán Calendar Wheel, *Codice Durán.*

Reprinted from *Codice Durán*. Electra y Tonatiuh Gutiérrez. Mexico: Arrendadora Internacional (1990), with permission.

Figure 3.16   Boban Calendar Wheel.

Reprinted from M. Veytia. [1907] (1973). *Los Calendarios Mexicanos*. Mexico: Museo Nacional. Reproduced courtesy of the John Carter Brown Library at Brown University.

may also be thought, as in the present case, to convey the passage of time.[21]

The Boban is a busy document that uniquely illustrates the intertwining of indigenous and European histories in the round. Two sets of drawings occupy the innermost portion of the quadripartite circle. In the top drawing, a pair of Texcocan town officials—one seated on a whirlpool in front of a temple, the other on a water mountain in front of a church—face each other. The figure on the right is identified as D. Hernando Pime[n]tel, an early colonial native official of Texcoco (Dibble 1990, p. 176). The speech glyphs that emanate from their mouths suggest that they are engaged in a dialogue. Kubler and Gibson (1951, p. 56) note that the 7 Tochtli date that appears in the text may be 1538, 1590, or 1642, the first being the best fit as it corresponds most closely to the chronology of the people who appear in the upper drawings. Glass (1975, p. 96) concurs. If this is the case the events referred to are not indicative of the later date when the drawing was composed (cf. Dibble 1990 for details).

In the bottom picture the two rulers of fifteenth-century Texcoco and Tenochtítlan, garbed in traditional clothing, replace the later historical figures. As is the case in the drawing above them, an indigenous image, a temple, flanks one personage, whereas a depiction of the Templo Mayor of Tenochtítlan is positioned behind the other. Below the lower figures, and separated by a text that has not been fully translated (a sun-faced disk is clearly part of it), a pair of Chichimec ancestors sit by a fire in front of a cave. The division of time into mythic (bottom) versus historic (top) is reminiscent of what one finds in the Mapa de Cuauhtinchan No. 2 (Carrasco and Sessions 2007) and similar documents from Central Mexico.[22]

Resumé and Discussion

The early colonial period was a time when theories of the organization of the solar system were hotly debated in scholarly circles. It also coincided with the most sweeping Old World calendar reform since the Roman Empire. Pope Gregory XIII implemented reform measures in 1582, among them a revised schedule of leap years. This latter historical coincidence may have been a source of the common misconception that the Aztec calendar employed leap years, a view still clung to by many today.[23] The friars who sought to interpret Central Mexican calendars would likely have been well aware of the scientific and theological debates surrounding these developments.

Failing to comprehend the nuances of the Mexica ceremonial cycle, the chroniclers attempted to codify whatever data they received about it in terms of structural patterns in the European calendar familiar to them, such as the 12 lunar months.[24] The *rotae* format common in Europe at the time became the obvious template for graphic representation of the passage of time.

What becomes clear in the present study thus far is that the earliest wheels from Central Mexico, for example, the Sahagún and Motolinía wheels, tend to focus strongly on computational aspects of the calendar. The chroniclers appear to have been motivated by a desire to understand and express how the Aztec calendar worked. On the other hand, many of the later calendar wheels seem to be more concerned with the business of placing religious holidays in the seasonal round. The emphasis on understanding the relationship between the 260- and 365-day cycles is also less evident in the later documents as the former cycle gradually disappears from those renditions. Specifically, as one moves from the earliest (1549) Motolinía wheel to the later copies of Valadés, Gemelli, and so on, the relative importance of the *tonalpohualli* diminishes; for example, it is the centerpiece in the Motolinía wheel, which includes only an accompanying text listing corresponding to *xihuitl* dates. In the (1579) Valadés wheel the month table occupies a larger share of space (the entire top half of the diagram). By 1697 the Gemelli wheel accords the month pictorials a central location accompanied by a matching six-part lunar cycle. This month-centered format is retained in all the later wheels. Note that the greater the distance in time forward from contact, the more dominant and the more embedded in the wheels become conventional European temporal conceptions, such as the four seasons and especially the month of the lunar phases, which, in some instances, seems to be force-fitted to the *xihuitl*.

In sum, colonial calendar wheels from Central Mexico variously misconceive and misrepresent indigenous ways of understanding and representing time. As we shall discover in chapter 5, following the discussion of Maya time wheels to follow in chapter 4, these colonial temporal instruments look not at all like what we see in the pre-Columbian codices. Heavily influenced by contemporary European debates surrounding the comprehension and expression of space/time, and by the extant circular cosmographic models used to represent these entities, the Spanish chroniclers appear to have created the calendrical models that they desired.

Figure 1.3 Calendar Wheels mesh in Codex Mexicanus, p. 9.

Codex Mexicanus, p. 9, from Central Mexico. Reprinted with permission of the Bibliothéque Nationale de France (BnE) (http://amoxcalli.org.mx/laminas.php?id=023-024&ord_lamina=023_09&act=con)

Figure 2.1a One example of Medieval and Renaissance wheel-shaped diagrams known as *rotae*. Copernican diagram of the solar system showing planets on circular orbs surrounded by the quintessential constellations of the zodiac.

Figure 3.2 European Wheel of solar declinations for successive dates of the seasonal year, 1550.

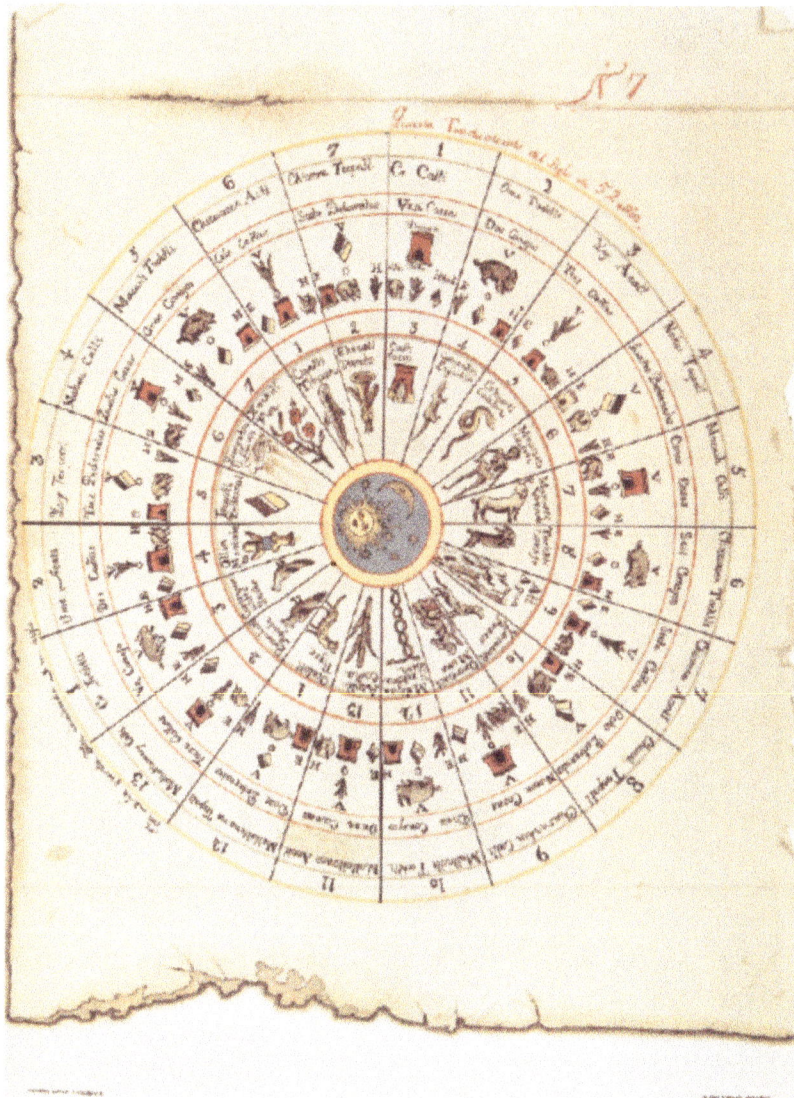

Figure 3.11   Veytia Wheel No. 7.

Reprinted from M. Veytia [1907] (1973). *Los Calendarios Mexicanos* Mexico: Museo Nacional.

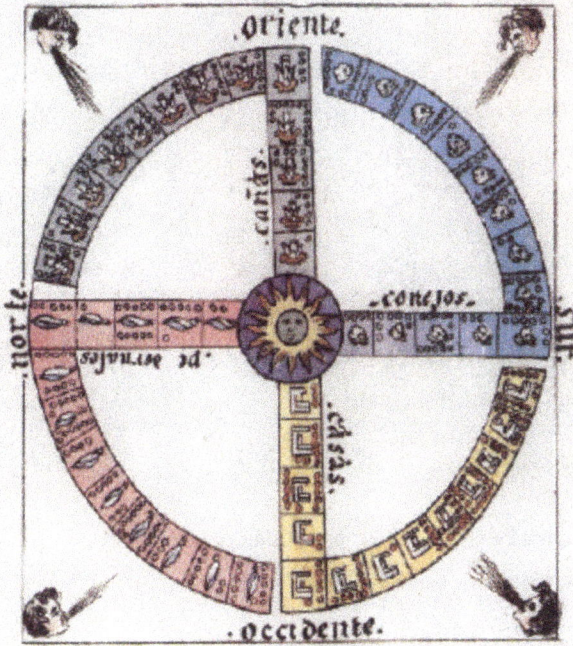

Figure 3.15 Durán Calendar Wheel, *Codice Durán.*

Figure 5.2 Codex Fejérváry-Mayer, p. 1. Codices Selecti, xxvi.

Reproduced with permission from Graz: Akad. Druck-u.Verlag.

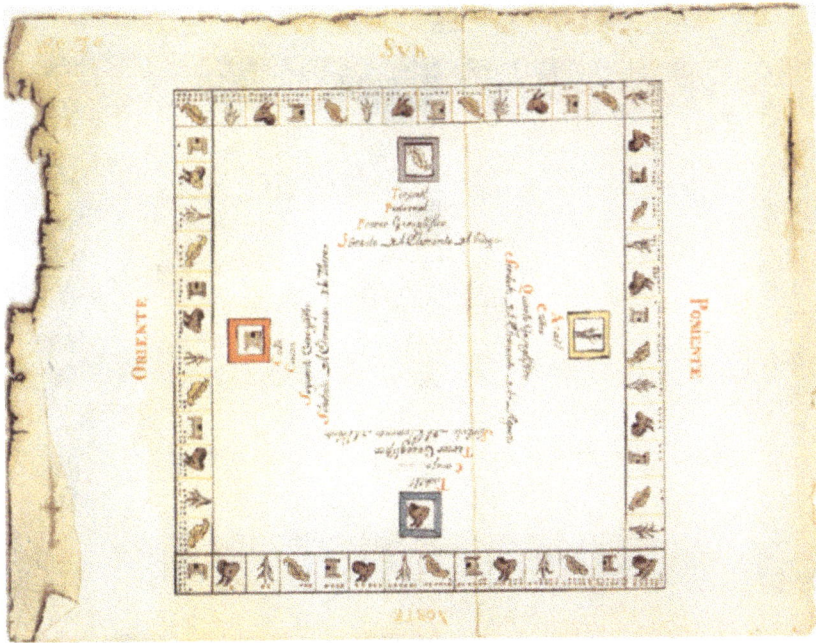

Figure 5.5 Veytia Calendar "Wheel" No. 3.

Reprinted from M. Veytia. [1907] (1973). *Los Calendarios Mexicanos* Mexico: Museo Nacional.

Figure 5.6 Aubin 1.

Reproduced with permission of the Bibliotheque Nationale de France (BnE), Paris.
(amoxcalli.org.mx/manuscritos analizados, no. 035-6)

IV

Maya Calendar Wheels

ONE OF THE MAJOR FEATURES OF COLONIAL Maya calendar wheels that sets them apart from their Central Mexican counterparts is that the content focuses most strongly not on the 365-day seasonal cycle but rather on scores of years 360 days in length, reckoned in the form of 360 x 20 = 7200-day periods known as *katuns*. On the other hand, and especially in the Postclassic Period, the year was always referenced via the year-bearer, the name and coefficient of the day in the 260-day count that ended a particular katun.[25] Given its greater number of cyclic components, the complex nature of the Maya calendar must have been considerably more confusing and perhaps even a bit intimidating to the Spanish chroniclers of Yucatan compared to those of Central Mexico. Glass's (1975) compendium lists seven Maya calendar wheels, all Yucatec, the four most important of which will be discussed here.

Landa's Katun Wheel

In a discussion of the calendar wheel he displays, Landa (Tozzer 1941, p. 168; Figure 4.1) says that Prehispanic Yucatec speakers

> had a certain way of counting the periods of time and their affairs by ages, which they did by periods of 20 years counting 13 twenties by means of one of the 20 letters of the months called Ahau, not in regular order but inverted as will be seen in the following circular diagram.

Note that Landa does not explicitly state that the Maya actually employed the circular format such as the one he uses for these computations.

The Landa wheel, the earliest of all known Maya examples (dated 1566) consists of 13 Ahau (one of twenty day names) faces arranged in radial compartments resembling a clock face. These are labeled in Roman numerals XIII, XI, IX, ..., as well as written out in Yucatec: *Oxlahun* [Thirteen] *Ahau, Buluc* [Eleven] *Ahau, Bolon* [Nine] *Ahau,* ..., reading clockwise from the ca. 11 o'clock position.[26] That Landa had trouble placing 13 entries on a conventional 12-unit clock face is

Figure 4.1 Landa's Katun Wheel.

Reprinted from A. Tozzer, ed. (1941). *Landa's Relación de las Cosas de Yucatan.* Cambridge: Papers of the Peabody Museum of Archaeology and Ethnology, Harvard, Vol. XVIII, p. 167.

evident. His artist seems to have made it to the 9 o'clock position before realizing that he would need to squeeze four entries into the remaining quadrant. A Christian cross on its side is positioned over XI Ahau at the top of the wheel; it begins the katun cycle. Landa calls the compartments "houses," which Tozzer (1941, p. 168, n. 884) likens to "the celestial home of the astrologer." The likeness intended may have been a reference to the Houses of the Greek horoscopic system, which are defined as 30 segments of the zodiac commencing at the eastern horizon and proceeding below it. An inscription at the center reads: "They call this count in their language Uazlazon Katun, that is to say the turning of the katuns."[27]

Calendar Wheel in the Book of Chilam Balam of Kaua

Miram and Bricker (1996) have noted that maps and diagrams in colonial reportorios bear a strong resemblance to European wind compass and zodiacal charts; for example, compare the wind compass on p. 127 of the Book of Chilam Balam of Kaua (Bricker and Miram 2002) (Figure 4.2) with the *rotae* in Figure 2.1, pp. 12, 14–15. Each of the 12-winged wind faces on the periphery of the circular frame of the Kaua puffs an animated breath toward the center of the circle.[28] Each wind is directionally labeled. Miram and Bricker speculate that six of the twelve wind directions refer to the place where the sun rises and sets on the solstices and equinoxes, thus implying a space–time connection in the diagram, though solar directions and temporal elements are not specifically referenced in the document. The calendar wheel on page 10 of the same document (Figure 4.3) exhibits a similar appearance. This quadripartite diagram, more appropriately a "katun wheel," is partitioned by an X, an arrangement in accord with the Mesoamerican concept of sides of space rather than cardinal points, to signify the four directions of the world. Milbrath (2001, p. 130) believes the signs at the end of the quadrant dividers may be Venus glyphs indicating the extreme horizon positions of that planet. The quartered circle representing the Earth at the center, on the other hand, has its four radials squared off with the layout of the page.

Thirteen curly-headed frontal faces, each labeled "Rey" ["King"], are pictured in one ring of the circle; these are labeled with coefficients 1–13. Here the artist appears to have dealt with the dilemma of equally dividing the number 13 four ways by squeezing a fourth face into the

Figure 4.2 Wind Compass in the Book of Chilam Balam of Kaua.

Reprinted from J. de Chavez. (1581). *Cronographía o Reportorio de los Tiempos, El Más Copioso y Preciso que Hasta Ahora Ha Salido a Luz.* Seville: Juan Francisco de Cisneros, folio 92.

left (north) side quadrant. On the other hand, the 20-day names are easily accommodated, five per quadrant. The four directions, to which the groups of days are assigned, are given Yucatec labels: east (*lakin*), north (*xaman*), west (*chikin*), and south (*nohol*). A Christian cross appears at the top (east side) of the diagram, as was customary in contemporary European maps, with the word *lakin* repeated. The names and numbers track the sequence of last days of a 20 x 360 = 7200-day period, which must end in a day named Ahau, with the coefficient receding by two per cycle as indicated previously (cf. note 26). Thus, the successive *katun* endings, named 13 (Ahau), 11 (Ahau), 9

Figure 4.3   Calendar Wheel in Book of Chilam Balam of Kaua.

Reprinted from V. Bricker and H. Miram. (2002). *An Encounter of Two Worlds: The Book of Chilam Balam of Kaua.* New Orleans: MARI Pub 68, p. 103. Reproduced courtesy of the Middle American Research Institute.

(Ahau) . . . , are read clockwise from the 12 o'clock position on the wheel.

Calendar Wheel in the Book of Chilam Balam of Chumayel

The thirteen crudely drawn Ahau faces that appear on the peripheral ring of this wheel (Figure 4.4), inscribed with Yucatec texts, are shown in a format translated by Roys (1933, p. 132) in Figure 4.5. The faces are positioned, one to each numbered compartment, with a fourteenth unlabeled compartment containing two faces added at the 1 o'clock position. Note the presence of a crescent moon and an effaced figure, possibly a second crescent, outside the fourteenth compartment, which are absent in Roys's version. Coefficients are spelled out in Yucatec inside the ring of faces and numbered in Arabic in the innermost ring.

Katun coefficients appear closer to the center of the circle; the sequence begins with 13 at the top and proceeds clockwise 11, 9 . . . ; the compartment containing the two Ahau faces between 13 and 11 are left blank, and the cross over 13 Ahau seems to indicate the end or the beginning of the cycle. Texts are written radially outside each compartment. Those over the blank compartment and compartments 5, 12, and 6 read, respectively, "to the east (south, west, north)," though they are not positioned in those directions on the page (assuming the Christian cross at the top of the diagram marks east, as was the convention); nor are they paired exactly opposite one another. Bowditch (1910, p. 330) thought the Maya intended to refer to horizon solstitial positions rather than intercardinals. He also attributed the curious division of the circle into 14 sectors to "ease of division."[29] As suggested earlier in the case of the Kaua, the chronicler may have failed to appreciate the Maya directional representation of *sides* or *quadrants of* rather than *points in* space. Unlike the other wheels, which are largely concerned with mechanics, divinatory information about each of the date entries is given. Thus, each katun carries with it a prophecy, for example, " . . . they shall be bent over and crippled . . . in the reign of Katun 5 Ahau," and so on.

Calendar Wheels in the *Book of Chilam Balam of Ixil*

This calendar wheel was referred to at the outset as one possible source of the contemporary representation of the Maya gear-wheel time machine (see Figure 1.2, p. 5). It consists of two parts: at the top a large

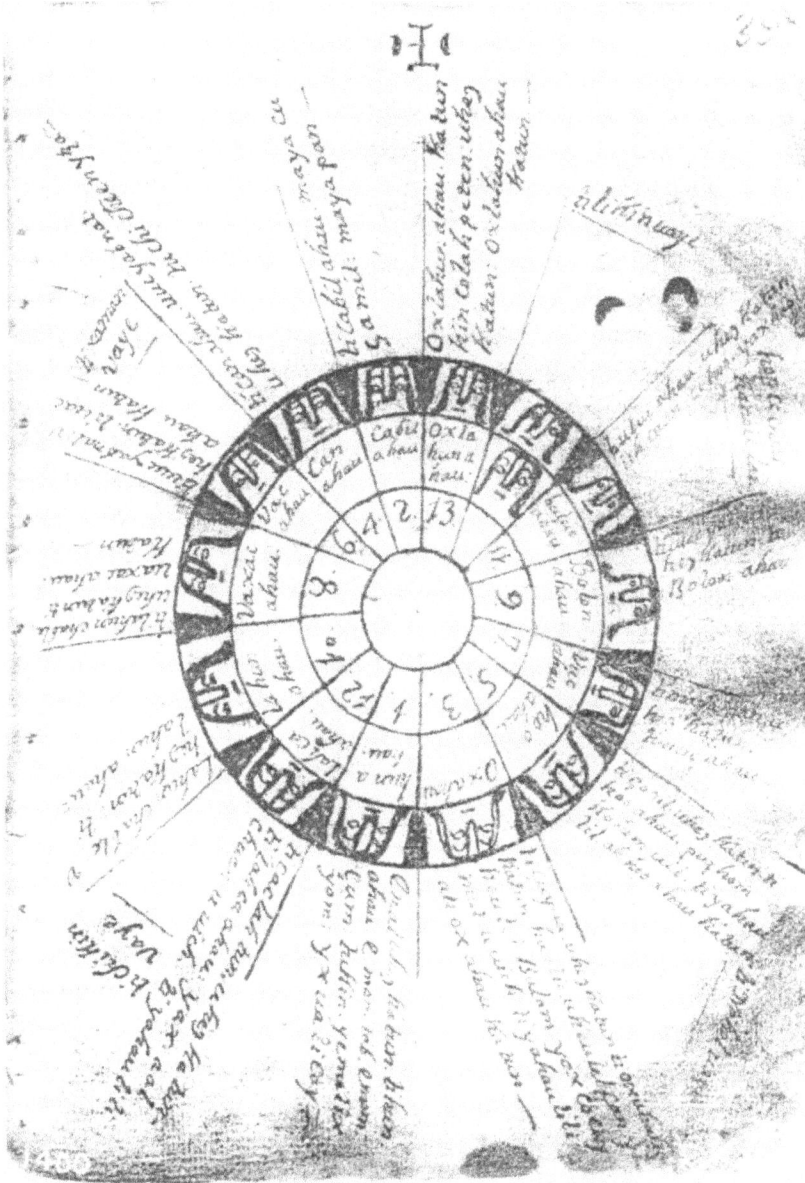

Figure 4.4   Chilam Balam of Chumayel Wheel.

Reprinted from G. Gordon. (1913). *The Book of Chilam Balam of Chumayel*. Philadelphia: University of Pennsylvania Museum Anthropological Publications No. 5, 11p. facsimile [1] 107, numb. 1. Reproduced courtesy of University of Pennsylvania Museum, image #14457.

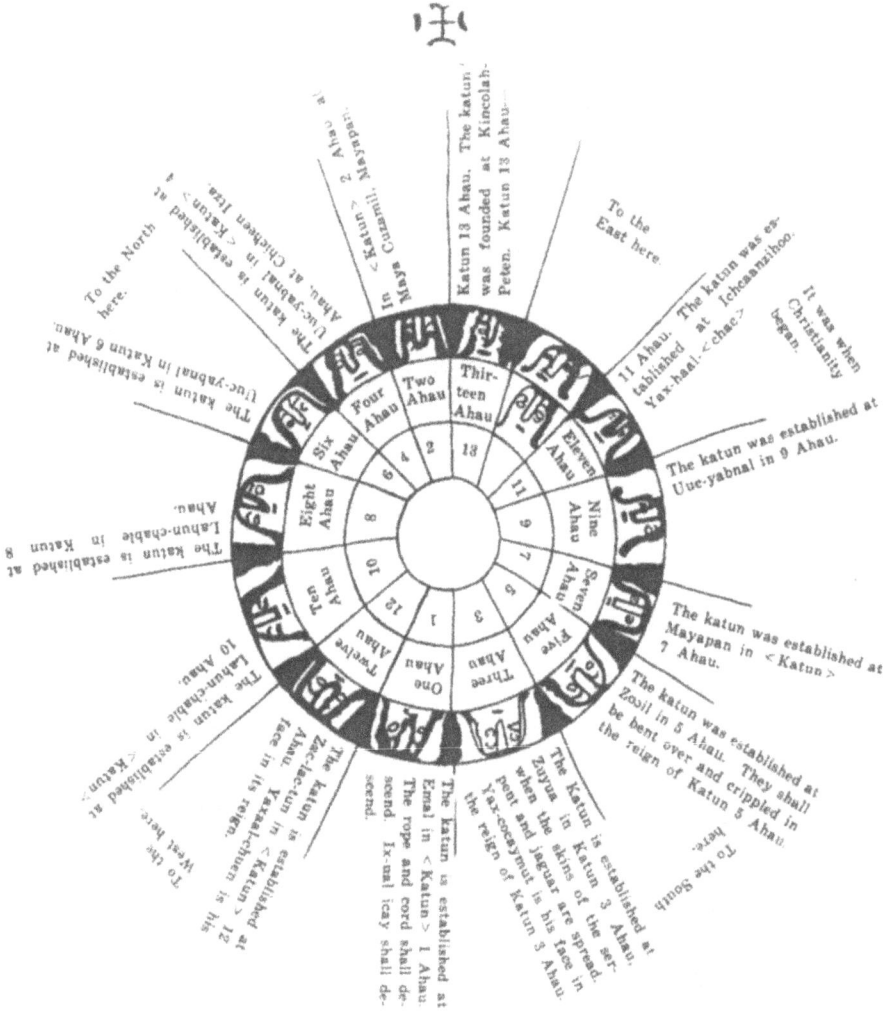

Figure 4.5 Translation of Chumayel Wheel.

Reprinted from R. Roys. (1933). *The Book of Chilam Balam of Chumayel*. Washington: Carnegie Institute of Washington, Pub. 438, p. 132.

wheel, labeled "*buk*," with 124 points on its periphery that distinctly resemble gear teeth; and at the bottom a smaller toothed wheel made up of 52 points. The top wheel shows 13 frontal faces, their numerical equivalents written in Mayan below, and the associated year-bearers in an outer circle. The outline of the face in the 12 o'clock position consists of a blank double circle, roughly executed; a vacant space at the 6 o'clock position again may reflect the problem of trying to create a symmetric clock face with 13 entries. Thus, the top wheel counts, from the 11 o'clock position and proceeding counterclockwise, the year-bearers spanning 12 years: *Hūkan, Muluc, Hiix, Cauac, Kaan, ...* The solid black circle at the center, which exhibits fiery-looking imagery extending from it, may have been intended to represent the sun. This flame or ray pattern distinctly resembles that found on the periphery of a circular diagram of the world in the Chavez reportorio (f. 96v).

The smaller wheel, labeled *"Acuch haab,"* displays year-bearer names and their corresponding directions surrounding a *croix formée*, which is further subdivided by intercardinals.[30] These are the same as Landa's year-bearers. Each is assigned a cardinal direction, with east (*lakin, kan*) at the top, south (*nohol, cauac*) at the right, west (*chikin, hiix*) at the bottom, and north (*xaman, muluc*) at the left. If we examine the reading direction, the likeness of this pair of wheels to the gear workings of a clock is evident. In the lower wheel the reading direction is clockwise. This means that a counterclockwise reading will result when its teeth mesh with those of the upper wheel, which would turn clockwise. This is in accord with the reverse reading order of the top wheel relative to the bottom. But given the number of gear teeth, especially the unusual number 124 indicated at the top, the two wheels, taken together, could not possibly have functioned as an instrument to mark time. (Recall that the same conclusion was reached in the case of the double wheel on Mexicanus 9, shown in Figure 1.3, p. 6.) Discussing the Ixil wheel, Perez had speculated that four "indications" of years equaled a *katun* and that there were "3 revolutions of the large wheel with the four hieroglyphs of the smaller one, making 12 signs, to which they added kan"—this would yield 13 years (cf. Bowditch 1910, pp. 328–9). *The Book of Chilam Balam of Mani* contains a copy of the Ixil wheel (Craine and Reindorp 1979, pp. 175–8). An accompanying table, written in native hand, which assumes knowledge of the continuous *tzolkin* count, explains how the day number can be found for any given day of any year.[31]

Resumé and Discussion

Referring to the Landa and Chumayel calendar wheels, Thompson
(1960, p. 47) noted: "I think these sequences of coefficients [of suc-
ceeding katuns] must have been illustrated by wheels, of which a few
colonial examples have survived," this despite Roys's opinion (Tozzer
1941, p. 167, n. 878) that "no known pre-Spanish representation of a
katun wheel or any other circular chronological diagram has yet been
found." On the other hand, Taube (1988) cites a carved limestone turtle
excavated in Str Q-244b at Mayapan as evidence to the contrary. The
outer rim of the oval-shaped shell pictures 13 contiguous Ahau signs
resembling what one finds on the Landa wheel. However, Miram and
Bricker (1996, p. 398) note that turtle carapaces are neither radially
symmetric, nor circular. Moreover, this item is not a calendar.[32] On the
other hand, Taube (2010, p. 213, fig. 24) also shows a turtle carapace
sectioned into quarters on a Classic period vessel from Kaminaljuyu.
Thirteen dots run along the inner periphery of the circle enclosing the
carapace. According to Taube, this represents the cosmic model of the
Earth floating in a sea with a solid celestial bowl capping it. Both Glass
and Brotherston argue that the circular shape of Central Mexican wheels
may have been influenced by that of the celebrated Aztec Sun Stone,
though, as mentioned earlier, this artifact also is not a calendar in any
functional sense (cf. Villela and Miller 2010, p. 3).

Although it would be incorrect to state that *no* representation of
time, functional or otherwise, in the Maya area prior to contact, is
circular, the present study leaves little doubt that the colonial Maya
katun wheels are heavily influenced by Western temporal views. There
is no convincing evidence that any of the wheels discussed herein,
whether Maya or Central Mexican/Tlaxcaltecan, were purely native in
origin; but native scribes, who had access to the European reportorios,
likely played an active role in the chroniclers' attempts to interpret
the nature of Mesoamerican time. Colonial calendar makers exploited
similarities between the native and Christian calendars, for example,
the resemblance between the concept of year-bearer and Dominical

Letter, the former giving the day of the *tzolkin* correspondence to New Year's Day, the latter having the day of the week correspond to New Year's Day. As far as the indigenous documents immediately preceding are concerned, the representation of the flow of time is almost exclusively not circular, but rather, as we shall demonstrate, quadrangular.

V

Indigenous Time in the Square

A T LEAST AT THE TIME OF CONTACT THE Maya concept of the universe was decidedly square, the world being thought to have the appearance of a rectangular house or *milpa* (cultivated field). According to translations of the creation story in the Popol Vuh (e.g., Tedlock 1985, p. 28), the dualistic male–female creator deity completed:

> the *fourfold* siding, fourfold cornering,
>
> measuring, fourfold staking,
>
> halving the cord, stretching the cord
>
> in the sky, on the earth,
>
> the four sides, the four corners, as it is said . . . . .

Contemporary ethnographic investigations support such a concept; for example, Sosa (1989, pp. 132–40) reports that the contemporary Yucatec Maya of Yalcobá, a town near Chichén Itzá, conceived of the four directions as sides of space: *u táan* (its face, its side), rather than the four cardinal points recognized in the West. The temporal element is incorporated into the spatial via the corners of the square, which are thought to be equivalent to the solstitial rise/set points. The square model of space/time is further replicated in the daykeeper's table or altar (cf. Sosa, see Figures 3.4 and 3.5, pp. 27, 29).

Of the 302 almanacs in the Post Classic Maya codices, not a single representation of the passage of time is laid out in a circular format. Dates of ceremonial offerings and the intervals between them are arranged linearly, in zigzag, and often in random patterns. Among pre-Columbian texts that show time in a rectangular form are the 4 x 5 rectangles listing all 20 day names on p. 32a of the Dresden Codex and p. 58c of the Madrid Codex. Larger *in extenso* almanacs (Just 2004) may be included in this category as well. Only the deer trapping and a few other almanacs convey even a vague resemblance to time in the round, dates and intervals being positioned at various points along the periphery of the body of the snared animal (Aveni 2004).

The same can be said of the Central Mexican precontact codices. Codex Borbonicus (pp. 21 and 22), the earliest colonial period calendrical indicator, is rendered in a square format (Milbrath 2001, p. 128). Also, time's course in the Postclassic Borgia Codex is delineated in linear bands in the form of a rectangle (cf. e.g., B.1–8, 61–70), a square (B.26), or a swastika (B.72). Page 25 of the Borgia Codex displays the day names of the *tonalpohualli* along four cardinal axes that radiate from a 10 Movement date at the center. The reading order is counterclockwise, moving to Flint, the innermost symbol at the left, then to Rain, the innermost day name at the bottom, and so forth. After one rotation the reader proceeds to the second set of four day names, then the third, and so on, thus creating a spiral path that moves outward.[33] Bricker's (2010, p. 327) rendition of the reading order is reproduced in Figure 5.1. The postcontact Dúran (see Figure 3.15, p. 51) and Tovar Wheels described previously exhibit a similar reading order, except that the insistence on the circular flow of time forces them to be rounded at the edges to fit the circular form of a swastika. It seems very likely that the indigenous spiral course that appears in Borgia 25 was still known when the layout of the Dúran wheel was set up. The fact that both the indigenous and the European format for displaying space-time are quadripartite would only have aided the transformation.

Other well-known examples of pre-Columbian calendrical layouts in the square format can be found in Codex Fejérváry-Mayer, page 1, a Mixtec–Puebla document (Figure 5.2; **see also color plate opposite p. 56**), and in the Madrid Codex, pp. 75–76 (Figure 5.3), a fifteenth-century document from west Yucatan. The basic format of these so-called "cosmograms" is the floral symbol with two sets of four petals: a "*croix formée*" with large trapezoidal petals and a "St. Andrew's Cross," or *Saltire*, consisting of four smaller rounded petals between the trapezoidal ones. A square design occupies the center. The *croix formée* shape of the cosmogram on Madrid pages 75–76 worked in much the same way.[34] The Madrid calendrical diagram demonstrates the willingness on the part of the Maya to borrow ideas from elsewhere, an attitude that persisted after the conquest with the importation of Spanish calendrical concepts. Each cosmogram is bordered by floral petals delineated with 260 circles. This ritual count is divided into cycles of 20 named days

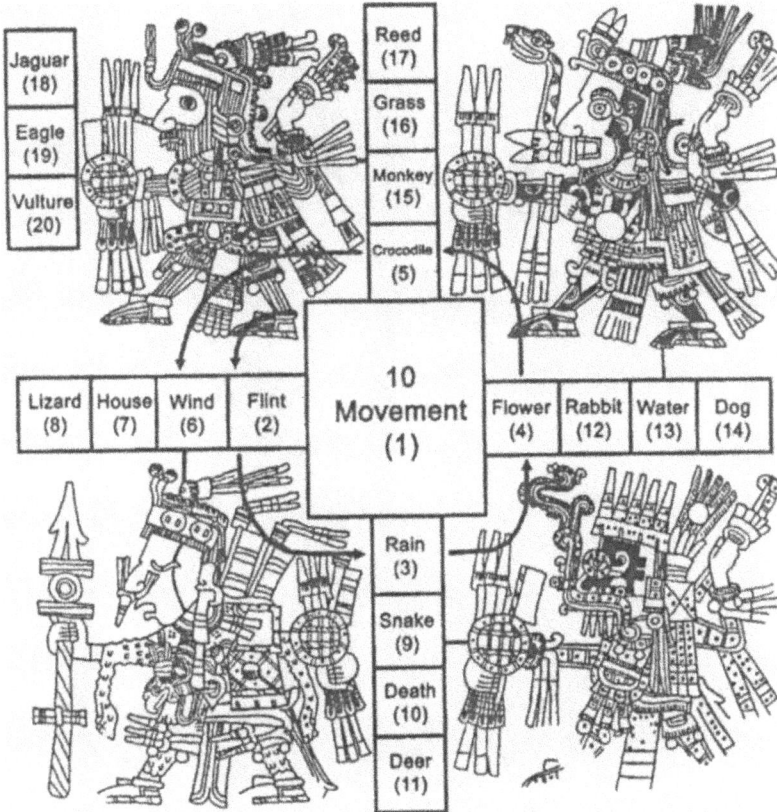

Figure 5.1   Diagram showing spiral course of the day-name sequence on page 25 of the Codex Borgia. Artwork by Christine Hernandez.

Reprinted from V. Bricker. (2010). "A Comparison of Venus Instruments in the Borgia and Madrid Codices." In *Astronomers, Scribes and Priests: Intellectual Interchange Between the Northern Maya Lowlands and Highland Mexico in the Late Postclassic Period*, G. Vail and C. Hernandez, eds. Boulder: University Press of Colorado, p. 327.

counted in groups of 13. In the Fejérváry, for example, the first set of 13 commences with 1 Cipactli (crocodile), whose teeth are visible just above the upper right-hand corner of the central square. Moving counterclockwise along the border, one proceeds to count 12 blue dots (shown on a dark field in the figure) completing the count of 13 on 1

Figure 5.2   Codex Fejérváry-Mayer, p. 1. Codices Selecti, xxvi.

Reproduced with permission from Graz: Akad. Druck-u.Verlag.

Ocelotl (jaguar), and so on, until the final segment of the cycle returns via Tochtli (rabbit) to 1 Cipactli. Thus the 260-day cycle is made to encapsulate all other astrological and calendrical matters depicted within the cosmogram (cf. Aveni 2001, pp. 149–51 for a detailed description).

Like Fejérváry 1, the early postconquest map of Tenochtítlan that appears on page 1v. of the Codex Mendoza exhibits a similar quadripartite fusion of space-time (Figure 5.4). Clearly rectangular in form, the map displays the eagle-cactus symbol at the center. This symbol is

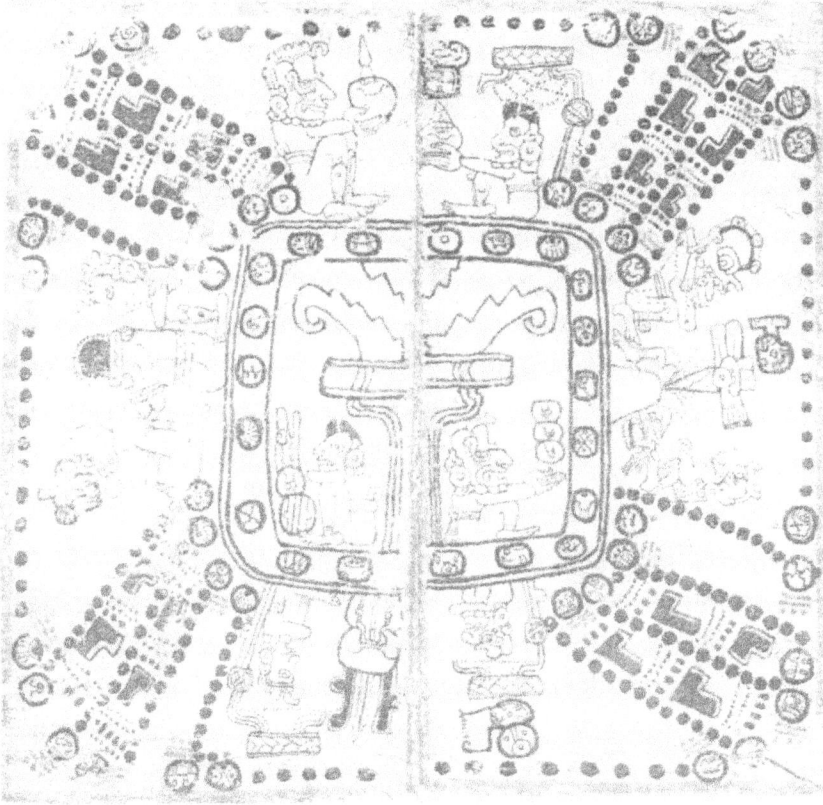

Figure 5.3 Codex Madrid.

Reprinted from J.A. Villacorta and C. Villacorta. (1977). *Códices Mayas*. Guatemala City: Tiopgrafía Naciona, pp. 75–76.

surrounded by a four-sided stream of blue water, with similar streams radiating outward from the center to the intercardinals. Ancestral rulers are pictured and named within the blue borders; below and outside them important battle/conquest scenes are shown. Also, as in Fejérváry 1, time flows in a (partial) rectangular border along the outer periphery. The 52-year count begins with year-bearer 1 Flint on folio 2v., then proceeds counterclockwise from 2 House (1325) at the upper left corner, to 3 Rabbit, and so forth, ending at 13 Reed, top center.[35] To appreciate just how different native perceptions of time and space were from those of the invader, one might contrast this Aztec-informed map of Tenochtítlan with Cortés's contemporary map of the city he had just conquered (cf. Mundy 1996, xii–xvii).

Veytia No. 3 (Figure 5.5; **see also color plate opposite p. 56**) is the only one in that collection of colonial documents that retains the square form. Like his other examples, it exhibits a curious mix of indigenous and colonial temporal concepts. The calendar consists of day symbols with coefficients in Aztec dot form. Each side of the square represents 13 years of a 52-year round. The calendar is read counterclockwise. The four year-bearer symbols are drawn inside the square: *Tecpatl* (south), *Calli* (east), *Tochtli* (north), and *Acatl* (west), which do not correspond to their usual directions. The content and linear form of Veytia No. 3 closely resemble that found in the Durán and Tovar wheels, with one significant difference: In the former one proceeds from a starting point all the way around the square in straight lines, turning at each corner, whereas in the latter the user jumps continuously from edge to edge, thus executing a spiral temporal course.

Each side of Veytia No. 3 is assigned an element, which is labeled in cursive along the boundary of a blank square interior space, from the east clockwise: Fire, Earth, Wind, and Water, respectively. These are the four basic elements delineated in the form of concentric circles in Aristotle's theory of natural place, though here they appear out of the usual order: Earth, Water, (Air), Fire, read from the center outward. Each year-bearer also exhibits a color direction: black, red, blue, and yellow, respectively. Glass (1975, p. 232) notes that no earlier version of this diagram is extant, but the calendar on p. 1 of Codex Aubin of 1576 (Figure 5.6; **see also color plate opposite p. 56**) emerges as the likely candidate from which Veytia No. 3 was copied. Unlike Fejérváry 1, which stands alone, it is attached to the annals history that follows it; specifically it depicts the 52 years that appear in the annals (Spitler 2005b, p. 281). Unlike Veytia No. 3, in the Aubin the 13 compartmental-ized sides appear detached from one another. Also, Aztec dots are replaced by Arabic numerals and a radiating sun face surrounded by the four year-bearers occupies the space at the center.

Figure 5.4   Mendoza Codex, p. 1.

Reproduced with permission from the Bodleian Library, Oxford University, MS Arch. Selden. A.1, folio 2r.

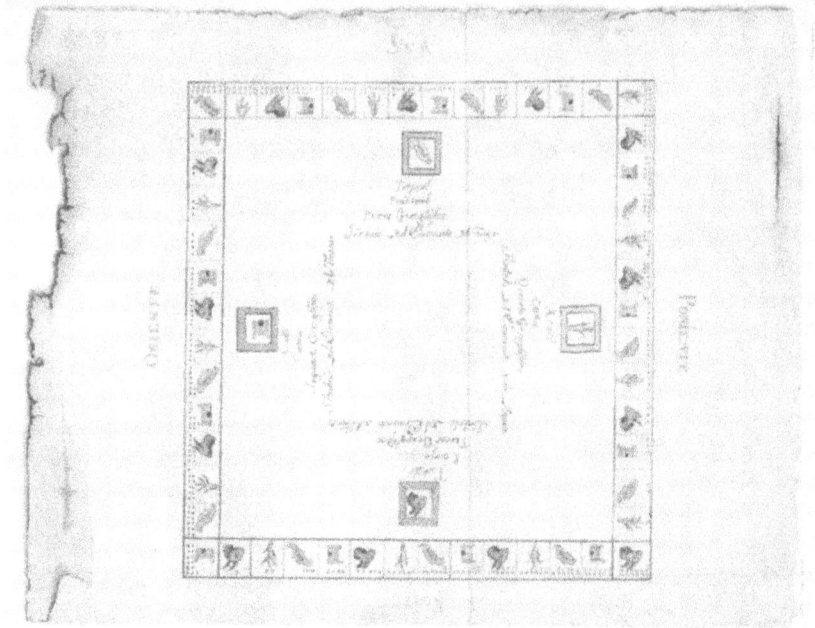

Figure 5.5 Veytia Calendar "Wheel" No. 3
Reprinted from M. Veytia. [1907] (1973. *Los Calendarios Mexicanos* Mexico: Museo Nacional

The temporal layout is reminiscent of the spatial distribution of *xiutonalli* ("year days" or year-bearers) about the Templo Mayor in colonial Aztec codices, for example, *Acatl, Tetamazolco* (east), *Tecpatl Necoquixecan* (north), *Calli, Atenchicalcan* (west), *Tochtli, Xolloco* (south). These *xiutonalli* represent the four entrances to the city that were visited in counterclockwise order during presacrificial ceremonies in the Aztec capital. The Spanish gloss attempts to make the 13-year divisions commensurate with three Olympiads (periods of four years), 12 being the most convenient European intervallic reference. Also the word "*casa*," the zodiacal term implying "House," is used to name each of the 13 days (Spitler 2005b, p. 281 and n.3).[30]

This study turned up a single exception to the rule of avoidance of the circle in precontact temporal formalisms in the codices. It appears

Figure 5.6   Aubin 1.

Reproduced with permission of the Bibliotheque Nationale de France (BnE), Paris.
(amoxcalli.org.mx/manuscritos analizados, no. 035-6)

in the Aubin ms. 20 (Figure 5.7). This late postclassic document features
a circle of 52 red dots surrounding the largely effaced imagery at the
center. Interestingly, the border of the rectangular diagram surrounding
it is delineated by 208 additional (red) dots, bringing the total to 260.
The space between the circle and the rectangle is occupied by four
scenes that represent, respectively, east (upper right), north (upper
left), west (lower left), and south (lower right); the effaced central scene
likely representing the fifth cardinal side of space. The scenes depict
various deities positioned on decorated altars, engaged in confrontation.
Boone (2007, p. 118) gives the reading order and identifies the deities
represented in each direction.[37] For a more detailed discussion of the
symbolism, see Jansen (1998).

Finally, it is worth noting that a number of pre-Columbian examples of time-in-the-round are found carved into the floors of buildings

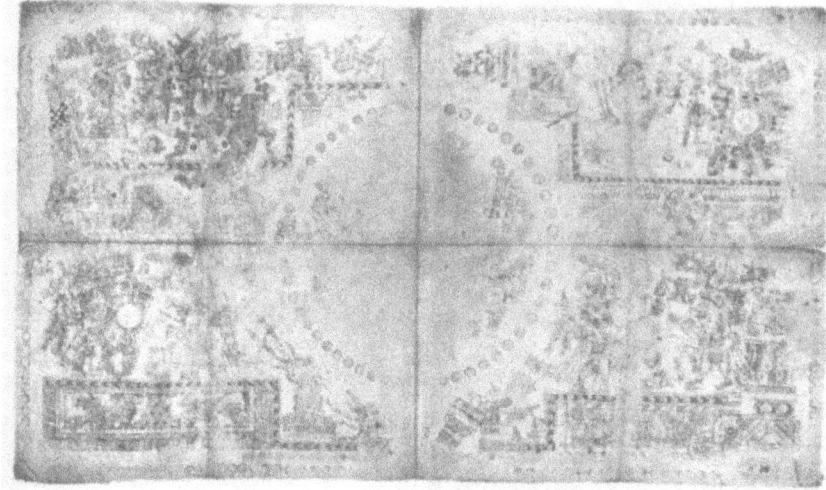

Figure 5.7   Aubin ms. 20.

Reproduced with permission of the Bibliotheque Nationale de France (BnE), Paris.

and on rock outcrops, particularly centered around Teotihuacan. These very early artifacts date from the middle of the first millennium A.D. Known as pecked crosses, they usually consist of double circles centered on a cross (Figure 5.8) Among the calendrically based subtallies and totals are 20 and 260. Moreover, one of the petroglyphs bears a distinct resemblance to the design of the Fejérváry and Madrid cosmograms, even down to the count of 260 elements along its outer periphery (Figure 5.9). Despite their remote antiquity, the circular form of these curious artifacts, of which some 75 examples exist, including one example from the Maya area (at the ruins of Uaxactún), cannot be completely ruled out as having had an influence on later representations of indigenous Mesoamerican time (cf. note 21). For a detailed discussion of pecked crosses, see Aveni (1989, 2000).

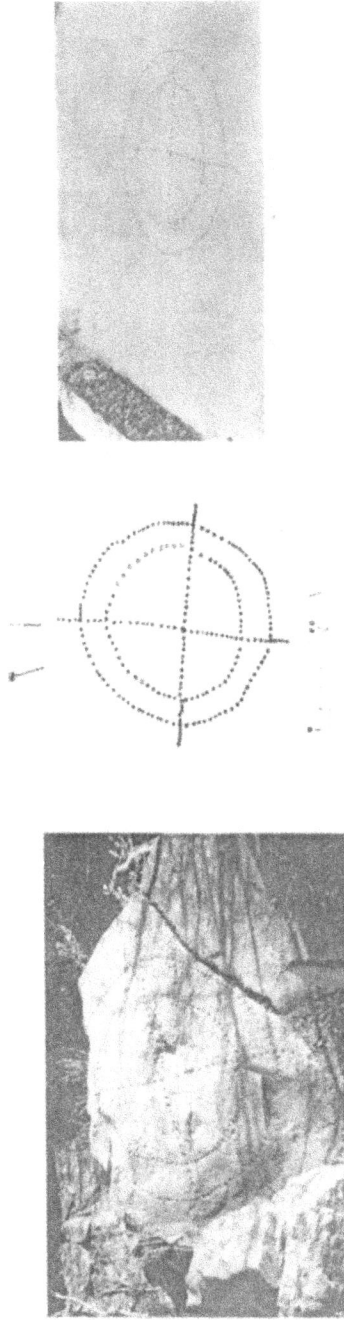

Figure 5.8   Pecked cross circles, early examples of pre-Columbian time in the round. Photographed by the author.

83

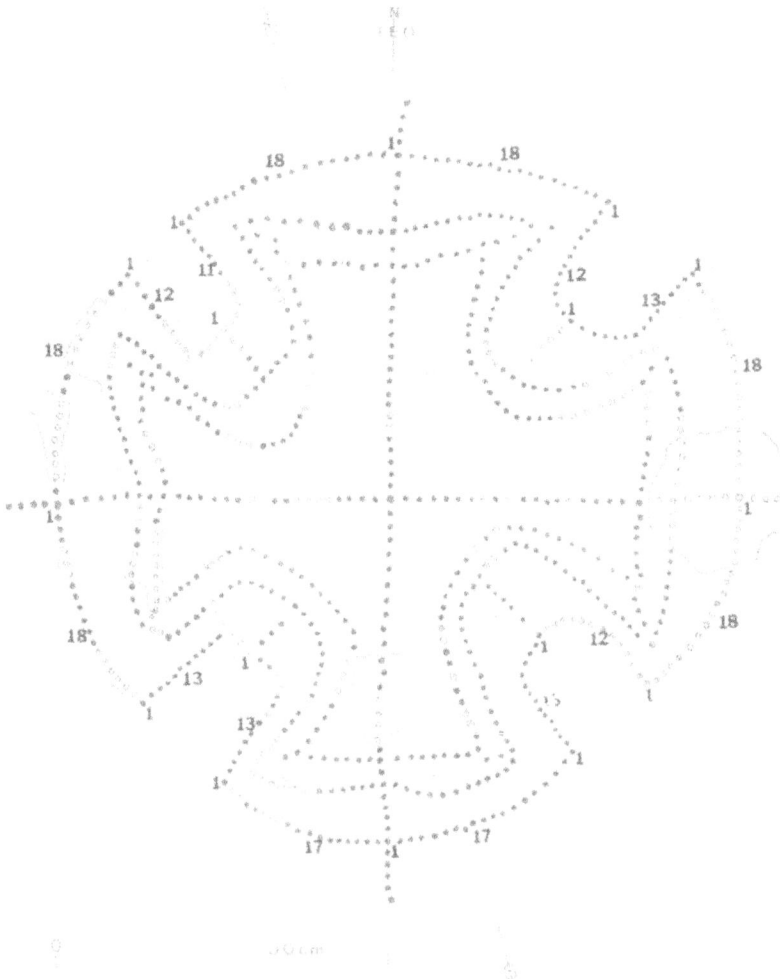

Figure 5.9 Pecked cross Teotihuacan No. 2 (cf. Figures 5.2 [p. 76] and 5.3 [p. 77]).
Photographed by the author.

VI

Summary

THE DEGREE OF EMBEDDEDNESS OF THE convention of the circle as a way of organizing and conveying knowledge of the world at the time of Hispanic contact cannot be overstated. The persistent imagery of wheels that grind out time and the perfectly round shape of the Platonic and Aristotelian universe remind us of both the gearwork on the inside and the face on the outside of a mechanical clock. Sonorous tones emanating from the clock's round, friendly countenance parsed out both sacred and secular time from every centrally located church or town hall in Renaissance Europe. But the form of the circle was further ingrained into the Renaissance mind in a much wider framework. *Rotae* served in many European books of knowledge as a means of lending order to the activities of the seasons, the arrangement of the houses and signs of the zodiac, even the ages of mankind. The representation of space conveyed in the *mappae mundi*, from the sixth-century B.C. geocentric world map of Hecataeus to Medieval T–O (orbis terrarum) maps, are as circular as the time's countenance mimicked on the face of the clock. The arrangement of matter in the cosmos found expression in circular shells and orbs in both the Ptolemaic geocentric and the Copernican heliocentric models of the universe, the sea change from one to the other having taken place shortly after the time of contact. So it seems logical that the native intellectuals who served as informants and assistants to the early chroniclers would have found it necessary to adapt to and make use of the circular form that dominated contemporary Hispanic ways of knowing and expressing the universe of space-time.

Landa, Motolinía, Sahagún, and Dúran demonstrate that circular diagrams became the preferred form of teaching and learning about the shape of Mesoamerican time. Such forms were both representational and computational, being concerned largely with the mechanics and hardly at all with the indigenous meaning of Mesoamerican time. The relatively few scholarly studies of these surviving documents concur that calendar wheels are an innovation of the colonial period (cf. Miram and Bricker 1993; Spitler 2005b, and the early documents such as the *Calendario indico* discussed earlier).

The present examination of Postclassic pre-Columbian calendrical documents indicates that both Maya and Central Mexican interpretations of time, especially the *tzolkin* (*tonalpohualli*), were expressed diagrammatically in square or rectangular form. Using circles to express the passage of Maya katuns (the so-called "katun round") seems to have worked quite well in colonial times, but, with a few exceptions, there is little evidence that in the pre-Columbian era these time units were conceived in such a form, nor were quadripartite divisions of katuns ever assigned to the cardinal directions (Bricker and Miram 2002, p. 75). Likewise, there is no reason to believe that the Central Mexican 365-day *xihuitl* was represented in circular form before the chroniclers (cf. also Russo 2005). Although our knowledge of the degree of access of the chroniclers to pre-Columbian documents is incomplete (e.g., many of the codices were hidden away and the few that are extant did not surface until well after the conquest), it is difficult to believe that the corpus of native documents could have been scantier in early colonial times than it is today. To judge from their comments, one suspects that the chroniclers did not bother to study these documents carefully at first, and only began to take their content seriously after they had destroyed most of them. When they did they resorted to a severely biased interview process.

Faced belatedly with the need to understand the mathematical correlations among time periods conceived in the native mentality, both for the sake of the conversion process and, for some, out of general interest in the problem, the Spanish chroniclers, assisted by their native informants, who must have been just as puzzled by the encounter between two vastly different ways of thinking about and representing time, relied very heavily on seeking likenesses between the native and European systems via the familiar circular forms of expression. Such a course of action further contributed to their failure to grasp many of the nuances of Mesoamerican timekeeping.

Consulting the pre-Columbian documents one finds instead that the square or the rectangle prevailed. This was a concept likely derived from the well-documented native concept of a quadripartite universe,

with directions regarded as borders or sides of space. This chosen form
was without a doubt the one to which the chroniclers would have been
exposed at the time of contact. They appear to have ignored it.

The data examined in this study reveal the nature of the intrusive
circular interpretation of Mesoamerican time. Details of the forced over-
lay of the European circular paradigm on indigenous temporality are
clearly reflected in several of the examples cited; for example, the
rounded spiral form of Motolinía's and Sahagún's calendar wheels likely
took their cues from native rectilinear diagrams, such as Borgia 25, that
were intended to be read in spiral fashion. The several attempts to
express the 365-day festival calendar in the form of a circle depicting
18 months of 20 days plus an additional five days likened the *xihuitl*
to the liturgical annual round in which festival dates (such as All Saint's
Day) were pinned down precisely to specific dates of the year. So too
was the trecena, or thirteen count, forced to resemble the hours on the
face of the ubiquitous European clock or the base-12 format of the
Western zodiac.[38] Other European concepts that intruded on native
forms of expressing space and time included the Greek elements, the
wind compass,[39] the synodic lunar month, the Olympiad, the four-
season year, the Jubilee Year and, not surprisingly given the concurrent
(late sixteenth century) European calendar reform, the concept of
leap years.

Notes

1. For example, we know that although young Nahua nobles were not taught astronomy, geometry, and arithmetic in the Franciscan school in Tlatelolco, the library contained books on these subjects, including a number of reportorios (Spitler 2005a, p. 97). Additionally, Lopez Austin (1973) has noted texts written in Nahuatl that contain material adapted from reportorios.

2. *Webster's New World Dictionary* 2003, New York: Simon and Schuster. The word "round" is suggestive.

3. The history of the life of Christ, on the other hand, is reenacted in a linear segment of a seasonal cycle.

4. Durán (Horcasitas and Heyden 1971, pp. 180–81) mentions a stone, most likely the round stone of Tizoc, that was unearthed on the site of the cathedral of Mexico City along with another pair, one of which was painted with the image of the sun, the other containing the "count of the years, months and days," but he says nothing about any of these being round.

5. The 260-day cycle, or *tonalpohualli*, consists of 20 day names paired with 13 numbers. Year-bearers are the *tonalpohualli* designations of New Year's Day in the 365-day calendar, or *xiuhmolpilli*. The two commensurate in a cycle of 52 *xiuhmolpilli*. Because 20 divides into 365 with a remainder of four, there can be only four year-bearers, each separated by five places in the 20-day name list, which repeat after four years. Their coefficients advance by one in successive years because 13 divided into 365 yields a remainder of one.

6. A closer look reveals where the concentrically drawn rings were sketched over to create the spiral effect (cf. between numbers 4 and 5 on the innermost ring).

7. This translation, and all others herein, unless otherwise stated, are attributed to E. Wake (personal communication, 11/12/10).

8. This translation was done by Susan Milbrath.

9. Specifically these are the last days of the 18 months of the indigenous calendar; the seven letters A–G, the dominical letters, repre-

sent the names of the days of the week that match these dates (for further discussion, see p. 48).

10. Derived from Hebrew tradition as a season of public rejoicing proclaimed every 50 years, in Roman Catholic tradition the Jubilee was at first scheduled every century before it was reduced to the shorter interval of 49 years. Additionally, note the crease (horizontal black line) that appears across the diagram at the division between the years 1558 and 1559, along which the document apparently had been folded.

11. On the other hand, Glass (1975, p. 231) believes Navas wrote the text in 1551.

12. Glass (1975, pp. 231–2 and Fig 74) traces the variants of this particular wheel.

13. This may be a reference to the notion that 13 days after a new moon, the full moon rises in the east after sunset. A more correct interval would be between 14 and 15 days.

14. This particular passage is, according to Gemelli, part of the "fábula" the Indians relate regarding the creation of the world at Teotihuácan.

15. The image of a snake consuming its own head may have been independently innovated in Dahomey, West Africa, as a symbol of kinship; cf. "The Nile Valley: African Civilizations History and Technology" (phppb_host.com). It may be worth noting as well that a pair of serpents circumscribe the Aztec sunstone.

16. Milbrath (personal communication, 3/12/12) speculates that the relation 360 divided by 6 approximates two lunar synodic months, which might have been intended to commensurate with three veintenas.

17. E. Wake (personal communication, 11/12/10) observes in it styles from different periods and places; thus the year-bearers, which do not align correctly, and lidded-eye star panels at the diagonals, look authentic, but are reminiscent of the style of Mixtec iconic script. Conversely, the month glyphs, with the exception of the five nemontemi,

are heavily Europeanized. Wake does not believe these are faces in profile, but they do carry strong parallels with the "hooked" equivalents on f.7r of the Codex Telleriano Remensis. Susan Milbrath (personal communication, 1/20/12) concurs on its status as a forgery, noting that the florid writing style is not characteristic of the 16th century. The Harvard, Veytia 5, Lorenzana, and Muñoz Camargo No. 1 exhibit similar script entries. Harvard and Veytia 5 give the same order of the months, though Muñoz No. 1 has Panquetzaliztli out of place. All include the *nemontemi*. The Veytia loses the *nemontemi* and gives different names to the fourth and fifth months. Oudijk (personal communication, 5/13/11) concurs with Wake that the PMAE wheel is a late falsification.

18. In most indigenous tropical cultures the seasonal division is twofold: wet and dry.

19. Another wheel (No. 1) is missing from all extant versions of the Serna manuscript (Glass 1975, p. 197)

20. Hebdomad implies the number seven, or a week. This is the Dibble and Anderson translation, with emendations by Oudijk (personal communication, 5/13/11).

21. The counts per quadrant are 19, 19, 19, and 16. This circular time count, along with the quadripartite circular form, is reminiscent of the sort of indigenous representation of time found in the Teotihuacan pecked crosses (Aveni 2000), which will be discussed later. Of footprints and the creation of the 20-day cycle by the first priest, the *Maya Book of Chilam Balam* (Roys 1993, pp. 116–7) states:

> What shall we say when we see man on the road?" These were their words as they marched along, when there was no man <as yet>. Then they arrived there in the east and began to speak. "Who has passed here? Here are footprints. Measure it off with your foot." So spoke the mistress of the world. Then he measured the footstep of our Lord, God, the Father. This was the reason it was called counting off the whole earth, *lahca* (12) Oc. This was the count, after it had been created by <the day> 13 Oc, after his feet were joined evenly, after they had departed there in the east.

22. For want of space a number of interesting wheels have been omitted from this study. Among them is a curious one dating from

1715 with astrological connotations (AGN Mapoteca, *inquisición* Vol. 778, pp. 277r–278r, Cat. No. 4877). The AGN catalogue description is as follows:

> Ilustración de un círculo astral, no queda claro quién lo realizó ni con qué motivo. Se observa un círculo grande y otro de 2.2 centímetros más angosto, en el espacio que se forma entre ambos se encuentran escritos numerous arábigos, se encuentra, a modo de plano cartesiano que atraviesa la circunferencia, una franja vertical que contiene numerous arábigos y en su parte inferior dice 'Media Muerte peor que malo y muy puesto', en la parte superior se encuentra escrito lo siguiente 'Media vida feliz aunque con dificultad'.

[Illustration of an astral circle ["horoscope"?]; it is unclear who prepared it or why. It consists of one large circle and a second, 2.2 cm. narrower. Arabic numerals are written in the space between them. In the manner of a Cartesian grid a vertical band containing Arabic numerals runs across the circumferences. Below are written the words "Death in part worse than bad, and imminent," and above "Life in part happy but with difficulties.] The wheel also exhibits a quadripartite form. At the tips and at the center of the wide band that divides the circle into four equal parts appears another quadripartite symbol: ※ Along a ring on the outer periphery and on the vertical portion of the band are listed, in seemingly random order, the numbers one to thirty, the approximate number of days in a lunar month of the phases. The first half of the count (1–15) is confined to the left half of the ring, whereas the second (16–30) occupies the right half; numbers from both counts share the vertical band. The partially deciphered text outside the diagram (clockwise from the right top) would appear to consist of omens. It reads: "Orta Maior," "Mors Minor," "Mors Maior," and "Orta Minor." As is typical in numerological divining (cf. Aveni 1996, pp. 63–66), a pair of tables assign numbers to alphabetic letters, thus A=1, B=2, . . . and the numbers 1–29 to numbers in that same range, though in no apparent order, thus 1=23, 2=14, ... Clearly, the wheel is of greater astrological than calendrical import, with some sort of horoscopic prediction relating to days of the lunar-phase cycle surely being intended.

Additionally, Oudijk and Castañeda de la Paz (2010) display a number of wheels in their work on the Boturini collection. Among the more unusual of these is a nine-compartmented wheel labeled "Rueda de los Nueve Señores de la Noche" [Wheel of the Nine Lords of the Night]. It consists of nine numbered, segmented compartments (read counterclockwise) in which are written the names of the Nine Lords of the Night. The figure of an owl occupies the center (Oudijk and Castañeda, fig. 7, #7). Another pair (figs. 5.1 and 6.4) stress the quadri-partition principle by employing spoke-like structures. In the latter case four rotated quadripartite year-bearer wheels appear as the featured elements.

23. For a review of possible ways the Aztecs might have reckoned the position of the xihuitl in the seasonal year, see Tena (1987).

24. In a little-known but highly criticized study, Brown (1977) theorized that the entire 18-month cycle was basically a European invention, which the friars thought of as a system of fixed and movable feasts patterned after the Christian seasonally based liturgical calendar.

25. It is not known how either a *haab* or a *tun* was named during the Classic Period. Scholars today name a Classic year—for example, when the day 0 Pop was a Caban day, that year (*haab*) was a Caban year. The day name, always Ahau, that ended a particular *katun* named that *katun* (without reference to year-bearers). The old orthography pertaining to Maya day names is employed throughout this book.

26. As 360 divided by 13 yields a remainder of 11, the *tzolkin* coefficient of each successive katun declines by two.

27. Brasseur pictures a copy of another wheel over his version of that of Landa. This wheel is divided into 12 compartments labeled Kan, Muluc, Ix, Cauac, Kan, . . . clockwise and numbered 1 next to Kan, 2 next to Muluc, . . . with 13 placed next to 1 in the same position as the initial Kan day at the top (Bowditch, 1910, p. 327, n. 1).

28. Actually the principal winds from the four cardinal directions blow toward the center; the faces representing them are frontal, whereas

the streams of air emanating from the adjacent faces (e.g., NE, NW), shown in profile, contribute their breaths to the "wind cloud" that emanates from the cardinal face; thus although there are eight wind faces, there are only four clouds, one in each of the cardinal directions.

29. Fourteen is also equal to the number of Articles of Faith and Works of Mercy in the religion of the conqueror.

30. Interestingly, Brasseur (reproduced in Bowditch 1910, fig. 62) had identified the central figure in his copy with the face of the moon, which he pictures with 30 decorative scallops. His drawing shows 137 points on the outer part of the wheel. Further, the faces in his drawing are evenly spaced and the blank face squeezed in at the top is absent.

31. As was the case with the Central Mexican calendar wheels, space does not permit presentation and extended discussion of further examples, which will only be referenced here with brief remarks. Thus, Weeks et al. (2009) have recently published a pair of calendar wheels in documents recorded by a resident nineteenth-century priest in the isolated highland Guatemalan town of Santa Catarina Ixtlahuácan. These are housed in the collection of the University of Pennsylvania library. Evidently the priest had acquired his information by interrogating a number of important native shamans in the area. The first document, dated 1722, is a 20-compartment wheel made with ruler and compass, with the *tzolkin* day names numbered, written out, and read clockwise beginning with 1 *Cauac*. The second, dated 1854, is essentially a replica of the first with nine added concentric rings outside of it. This device appears in two parts. The 180 compartments of each part list the days of the Christian year, beginning May 1. By reading clockwise in spiral fashion one can look up the day of the seasonal year and find the matching *tzolkin* entry or vice versa; for example, the 1 Cauac days for the year 1854 turn out to be May 3, May 23, Jun 12, ...

Another calendar wheel, this in the form of a mural painting, has recently been recovered in the upper cloister of a Franciscan monastery in Motul, Yucatan (www.colonial-mexico.com/Yucatan/yucmur als.html). Dated to the late eighteenth century, its clock-like form is set

in a square frame consisting of elaborate floral decorations. Personified
winds are pictured at the corners of the square. The eight sectors of
the wheel are subdivided in three to make a total of 24. The months
listed around the outside each occupy two sectors. A white star with
a long star-studded tail marks the center. The head, hands, and feet of a
humanoid figure that holds the frame mark the solstices and equinoxes,
whereas the four winds denote the halfway points in between.

32. The same can be said of the calendrical glyphs executed in a
roundish form on Medallions 3 and 4 found in Structure I-SUB of
Dzibilchaltun (Garcia Campillo and Lacadena 1992.)

33. An expression of the indigenous conception of time in spiral
form, though not expressed graphically, may be evident in the well-
known *voladores* ceremony in which four "flyers" circulate around a
pole thirteen times each to commemorate the 4 x 13 = 52-year calendar round.

34. Interestingly, in both examples a disordered pattern for time
reckoning occurs at the center. In the Madrid case it involves an embedded Venus calendar, whereas in the Fejérváry example it concerns the
Lords of the Night (cf. Bricker 2010, fig. 12, Boone 2007, fig. 65).

35. The last page of the Princeton Catechism (*An Otomí Catechism*
1968) appears to be a copy of the rectangular/cross pattern on Mendoza 1v.

36. Other versions of Aubin 1 display variations. For example, in
the 1836 edition of Veytia the four year-bearer glyphs (uncaptioned)
are pictured in quarters of a square at the center of the diagram. Lines
radiating from each of them partition the square border into 4 x 13
numbered compartments containing the day-name symbols. The count
begins not at the corner but in the middle of each side. There is no
sun face. In another version found in the Boturini collection (Oudijk
and Castañeda de la Paz, 2010, fig. 5) the sun face at the center is
rotated 90 counterclockwise. An accompanying diagram shows four

wheels, each at the end of a spoke of a quadripartite form, with rotating year-bearers.

37. Boone's version of the count begins at the midpoint of the right side of the rectangle, passes upward, then turns left, running across the top and then down the left side to its midpoint. Having reached 104, the reader moves inward to the left side of the central circle and counts 52 additional red dots clockwise before exiting and resuming the clockwise count at the left midpoint of outer rectangle, ending back at the right midpoint, where the total reaches 260.

38. The Maya did possess a zodiac of their own, though it was not expressed in a circular format. It consisted of 13 constellations and may have operated in a different manner from the one known in the West (cf. Aveni 2001, pp. 200–4 for an overview). There is no evidence that refers to this zodiac in any of the documents reviewed here.

39. Miram and Bricker (1996) make a strong case that the wind compass, another European device that connected space and time, also served as a graphic scheme that influenced the interpretation of the Maya calendar. The same may hold for Central Mexico. They note that the 12 points of the compass not only resonate with the zodiacal signs but also six of them can be related to key positions of the sun at the horizon (the solstices and the equinoxes).

References

Acuña, R., ed. 1984. *Relaciones Geográficas del Siglo XVI, Vol. 4, Tlaxcala I* (Muñoz Camargo, Codex Tlaxcala). Mexico: UNAM.

Acuña, R., ed. 1986. *Relaciones Geográficas del Siglo XVI, Vol. 7, Mexico II.* Mexico: UNAM.

An Otomí Cathechism. Introduction by G. Griffin. 1968. Princeton and Meriden, CT: Meriden Gravure Co.

Aubin, J. 1893. "Histoire de la Nation Mexicana depuis le départ d'Aztlan jusqu'à l'arrivée des Conquerants Espagnols (et au delà 1607). Ms. figurative accompagné de texte en Langue Nahuatl ou Mexicaine Sulvi d'une Traduction en Francaise par feu. JMA Aubin Reproduction of the Codex de 2576 (1576). Paris.

Aveni, A. 1989. "Pecked Cross Petroglyphs at Xihuingo." *Archaeoastronomy* 14, JHA, xx: S73–S115.

———. 1996. *Behind the Crystal Ball: Magic, Science, and the Occult from Antiquity through the New Age.* New York: Random House.

———. 2000. "Out of Teotihuacan: Origins of the Celestial Canon in Mesoamerica." In *Mesoamerica's Classical Heritage: From Teotihuacan to the Aztecs,* D. Carrasco, L. Jones, and S. Sessions, eds., 253–68. Niwot: University of Colorado Press.

———. 2001. *Skywatchers: A Revised and Updated Version of Skywatchers of Ancient Mexico.* Austin: University of Texas Press.

———. 2004. "Intervallic Structure in Cognate Almanacs in the Madrid and Dresden Codices." In *The Madrid Codex: New Approaches to Understanding an Ancient Maya Manuscript,* G. Vail and A. Aveni, eds., 131–46. Boulder: University Press of Colorado.

Bartholomaeus Anglicus, 1485. *De Proprietatibus Rerum.* Lyons.

Baudot, G. 1995. *Utopia and the History of Mexico.* Niwot: University Press of Colorado.

Beer, A., and P. Beer, eds. 1975. *Kepler: Four Hundred Years.* Oxford: Pergamon.

Berthe, J.P. 1968. *Le Mexique à la Fin du XVIIᵉ Siecle vu par un Voyageur Italien, Gemelli-Careri.* Paris: Callman-Levy.

Boone, E. 2007. *Cycles of Time and Meaning in the Mexican Books of Fate.* Austin: University of Texas Press.

Bowditch, C. 1910. *The Numeration, Calendar Systems and Astronomical Knowledge of the Mayas.* Cambridge, Mass.: Cambridge University Press.

Bricker, V. 2010. "A Comparison of Venus Instruments in the Borgia and Madrid Codices." In *Astronomers, Scribes and Priests: Intellectual Interchange Between the Northern Maya Lowlands and Highland Mexico in the Late Postclassic Period,* G. Vail and C. Hernandez, eds., 309–32. Boulder: University Press of Colorado.

———, and H. Miram. 2002. *An Encounter of Two Worlds: The Book of Chilam Balam of Kaua.* New Orleans: MARI Pub 68.

Brotherston, G. 2005. *Feather Crown: The 18 Feasts of the Mexican Year.* Oxford: Oxbow.

Brown, B. 1977. "European Influences in Early Colonial Descriptions and Illustrations of the Mexica Monthly Calendar" (Ph.D. dissertation). Albuquerque: University of New Mexico, Art History Department.

Carrasco, D., and S. Sessions, eds. 2007. *Cave, City, and Eagle's Nest: An Interpretive Journey Through the Mapa de Cuauhtiachan No. 2.* Albuquerque: University of New Mexico Press.

Chavez, J. de 1581. *Cronographía o Reportorio de los Tiempos, El Más Copioso y Preciso que Hasta Ahora Ha Salido a Luz.* Seville: Juan Francisco de Cisneros.

Craine, E., and R. Reindorp. 1979. *The Codex Pérez and the Book of Chilam Balam of Mani.* Norman: University of Oklahoma Press.

Cuesta Domingo, M. 1998. *La Obra Cosmográfica y Náutica de Pedro de Medina.* Madrid: BCH.

D'Olwer, L. 1987. *Fray Bernardino de Sahagún 1499–1590.* Salt Lake City: University of Utah Press.

Dibble, C. 1990. "The Boban Calendar Wheel." *Estudios de Cultura Nahuatl* 20: 173–82.

Dyer, N. 1996. *Fray Toribio de Benavente Motolinía Memoriales.* Mexico: El Colegío de Mexico.

Edmonson, M., ed. 1974. *Sixteenth Century Mexico: The Work of Sahagún.* Albuquerque: University of New Mexico Press.

Garcia Campillo, J., and A. Lacadena G. 1992. "Sobre Dos Textos Glificos del Postclásico de Dzibilchaltun." *Mayab* 8: 46–53.

Garcia Cook, A. 1973. "Algunos descubrimientos en Tlalancaleca', Edo. de Puebla." In *Comunicaciones. Proyecto Puebla-Tlaxcala,* 9. Puebla: Fundación Alemana para la Investigación Científica.

Garcia Icazbalceta, J. 1892. *Nueva Colleción de Documentos para la Historia de Mexico,* Vol. 5 (Codice Mendieta).

Gingerich, O. 2002. *An Annotated Census of Copernicus'* De Revolutionibus (Nuremberg, 1543 and Basel, 1566). Leiden: Brill.

Glass, J. 1975. "A Census of Native Middle American Pictorial Manuscripts." In *Guide to Ethnohistorical Sources,* H. Cline, ed. Part 3: *Handbook of Middle American Indians 14,* R. Wauchope, ed., 81–252. Austin: University of Texas Press.

Gordon, G. 1913. *The Book of Chilam Balam of Chumayel.* Philadelphia: University of Pennsylvania Museum Anthropological Publications, No. 5.

Heath, T. 1932. *Greek Astronomy.* London: Dent.

Horcasitas, F., and D. Heyden. 1971. *Book of the Gods and Rites and the Ancient Calendar by F. Diego Durán.* Norman: University of Oklahoma Press.

Jansen, M. 1998. "La Fuerza de los Cuatro Vientos. Los Manuscritos 20 y 21 del 'Fonds Mexicain'". *Journal de la Societé des Américanistes* 84(2): 125–61.

Just, B. 2004. "*In Extenso* Almanacs in the Madrid Codex." In *The Madrid Codex: New Approaches to Understanding an Ancient Maya Manuscript,* G. Vail and A. Aveni, eds., 225–76. Boulder: University Press of Colorado.

Kline, N.R. 2001. *Maps of Medieval Thought: The Hereford Paradigm.* Woodbridge: Boydell.

Koestler, A. 1959. *The Sleepwalkers.* New York: MacMillan.

Kristeller, P. 1978. "The First Printed Edition of Plato's Works and the Date of Its Publication (1484)." In *Studi Copernicana: Science and History,* E. Hilfstein, P. Czartoryski, and F. Grande, eds., 25–35. Warsaw: Polish Academy of Science.

Kubler, G., and C. Gibson. 1951. "The Tovar Calendar: An Illustrated Mexican Manuscript, ca. 1585." New Haven: Memoirs of the Connecticut Academy of Sciences.

Lopez Austin, A. 1973. "Un Repertorio [sic] de los Tiempos en Idioma Nahuatl" *Anales de Antropologia* 10: 285–96.

————. 1974. "The Research Method of the Questionnaires." In *Sixteenth Century Mexico: The Work of Fray Bernardino de Sahagún*, M. Edmonson, ed., 111–49. Albuquerque: University of New Mexico Press.

Lorenzana, F. 1770. *Historia de Nueva-España, Escrita por su Eslarecido Conquistador Hernán Cortés, Aumentada con Otros Documentos y Notas*. Mexico.

Martinez, H. (1606) 1991. *Reportorio de los Tiempos e Historia Natural de Esta Nueva España*, F. de la Maza, ed. Mexico: Consejo Nacional Bara la Cultura y las Artes.

Mathes, M. 1982. Santa Cruz de Tlatelolco: La Primera Biblioteca Académica de las Américas. Mexico: Archivo Historico Diplomático Mexicano. Secretária de Relaciónes Exteriores.

Milbrath, S. 2001. "Calendar Wheels." *Oxford Encyclopedia of Mesoamerican Cultures*, Vol. 2, D. Carrasco, ed., 128–30. Oxford: Oxford University Press.

————. 2011. *Heavenly History: Ancient Mexican Astronomy in the Codex Borgia*. Austin: University of Texas Press.

Miram, H., and V. Bricker. 1996. "Relating Time to Space: The Maya Calendar Compasses." In *Palenque Round Table* (1993) Vol. X, M. Greene Roberston, M. Macri, and J. McHargue, eds., 393–402. San Francisco: Precolumbian Art Research Institute.

Mundy, B. 1996. *The Mapping of New Spain: Indigenous Cartography and the Maps of the Relaciones Geográficas*. Chicago: University of Chicago Press.

Murdock, J. 1984. *Album of Science*. New York: Scribners.

Noguera, E. 1964. "El Sarcófago de Tlalancaleca". *Cuadernos Americanos* 3: 139–48.

Oudijk, M., and M. Castañeda de la Paz, 2010. "La Colección de Manuscritos de Boturini: Una Mirada desde el Siglo XXI". In *El Caballero*

Lorenzo Boturini: Entre dos Mundos y dos Historias, M. Oudijk and M. Castañeda de la Paz, eds., 87–128. Mexico: Museo de la Basílica de Guadalupe.

Perujo, J. 1976. *G.F. Gemelli Careri Viaje a la Nueva España*. Mexico: UNAM.

Ramirez, J.F. 1898. Adiciones a la Biblioteca de Beristáin, Mexico: Agüeros (Biblioteca de Autores Mexicanos, v. 16–17).

————. 2001–2003. *Obras Historicas*, E. de la Torre Villar, ed. Mexico: UNAM (4 vols.).

Robertson, D. 1959. *Mexican Manuscript Painting of the Early Colonial Period: The Metropolitan Schools*. New Haven: Yale University Historical Publications.

————. 1966. "The Sixteenth Century Encyclopedia of Fray Bernardino de Sahagún." *Cahiers d'Histoire Mondiale* 9(3): 617–27.

Roys, R. 1933. *The Book of Chilam Balam of Chumayel*. Washington: Carnegie Institute of Washington, Pub. 438.

Russo, A. 2005. *El Realismo Circular: Tierras, Espacios y Paisajes de la Cartografía Indígena Novohispania Siglos XVI y SVII*. Mexico: UNAM.

Sahagún, B. de 1953. *Florentine Codex, Book 7*, A.J.O. Anderson and C.E. Dibble, eds. and trans. Santa Fe: SAR and University of Utah Press.

————. 1979. *Florentine Codex, Books 4 and 5*, A.J.O. Anderson and C.E. Dibble, eds. and trans. Santa Fe: SAR and University of Utah.

Serna, J. de la (1654) 1892."Manual de Ministros de Indíos para el Conocimiento de sus Idolatrias, y Extirpación de Ellas." In *Tratado de las Supersticiones, Dioses Ritos, Hechicerías, y Otras Costumbres Gentilicas de las Razas Aborigenes de Mexico*, 47–368. Mexico: Ed. Fuente Cultural.

Sosa, J. 1989. "Cosmological, Symbolic, and Cultural Complexity Among the Contemporary Maya of Yucatan." In *World Archaeoastronomy*, A. Aveni, ed., 130–42. Cambridge, UK: Cambridge University Press.

Spitler, S. 2005a. *Nahua Intellectual Responses to the Spanish: The Incorporation of European Ideas into the Central Mexican Calendar* (Ph.D. Dissertation). New Orleans: Tulane University.

————. 2005b. "Colonial Mexican Calendar Wheels: Cultural Translation and the Problem of Authenticity." In *Painted Books and
Indigenous Knowledge in Mesaomerica: Manuscript Studies in Honor
of Mary Elizabeth Smith*, 271–88. New Orleans: Middle American
Research Institute, Pub. 69.

Steck, F. 1951. *Motolinía's History of the Indians of Spain*. Washington:
Academy of American Franciscan History.

Stuart, G. 1975. "The Maya Riddle of the Glyphs." *National Geographic*
148(6): 768–91.

Taube, K. 1988. "A Prehispanic Maya Katun Wheel." *Journal of Anthropological Research* 44(2): 183–03.

————. 2010. "Where Earth and Sky Meet: The Sea and the Sky in
Ancient and Contemporary Maya Cosmology." In *Fire Pool: The
Maya and the Mythic Sea*, D. Finamore and S. Houston, eds., 202–19.
Salem, MA: Peabody Essex Museum.

Tedlock, D. 1985. *Popol Vuh: The Mayan Book of the Dawn of Life*. New
York: Simon and Schuster.

Teeple, J.E. 1930. "Maya Astronomy." *Contributions to American Archaeology* 1: 29–115.

Tena, R. 1987. *El Calendario Mexica y la Cronografía*. Mexico City:
INAH.

Thompson, J.E.S. 1960. *Maya Hieroglyphic Writing*. Norman: University
of Oklahoma Press.

Tozzer, A., ed. 1941. *Landa's Relación de las Cosas de Yucatan*. Cambridge: Papers of the Peabody Museum of Archaeology and Ethnology, Harvard, Vol. XVIII.

Valades, D. de (1579) 1989. *Retorica Cristiana*, facsimile edition. Mexico: FCE.

Veytia, M. (1907) 1973. *Los Calendarios Mexicanos* Mexico: Museo
Nacional.

Villacorta, J.A., and C. Villacorta. 1977. *Códices Mayas*. Guatemala City:
Tiopgrafía Nacional.

Villela, K., and M. Miller, eds. *The Aztec Calendar Stone*. Los Angeles:
Getty Publications.

Wake, E. 2010. *Framing the Sacred: The Indian Churches of Colonial Mexico*. Norman: University of Oklahoma Press.

Weeks, J., F. Sachse, and C. Prager. 2009. *Maya Daykeeping: Three Calendars from Highland Guatemala*. Boulder: University Press of Colorado.

Wilkerson, J. 1974. "The Ethnographic Works of Andrés de Olmos." In *Sixteenth Century Mexico*, M. Edmonson, ed., 27–77. Albuquerque: University of New Mexico Press.

Zamorano, R. 1585. *Cronología y Reportorio de la Razón de los Tiempos*. Seville: Andrea Pescioni y Juan de Leon.

Index

A

Acuña, R., 23, 44
Ages of Man, 13
Ajq'ijes, 8
Anglicus, Bartholomaeus, 13
Annals of the Cakchiquels, 8
Antiguas, Pinturas, 44
Aristotle, 11, 46, 78
Austin, Lopez, 45
Aveni, A., 28, 32, 73, 76, 82, 95–96,
    100

B

Baudot, G., 24–25
Bede, Venerable, 13
Benavente, Fray Joan, de, 36
Boban Calendar, 50–53
Book of Chilam Balam of Chumayel,
    63–64
Book of Chilam Balam of Ixil, 4, 5,
    64–69
Book of Chilam Balam of Kaua, 8,
    61–63
The Book of Chilam Balam of Mani,
    67
Books of Chilam Balam, 8
Boone, E., 81
Borgia Codex, 74
Bowditch, C., 64, 67
Bricker, V., 8, 61, 74, 88
Brotherston, G., 28, 47, 49
Burning of Maya codices, 8

C

Calendar keepers, 8
Calendar Wheel of Motolinía, 2

Calli xihuitl, 19
Carrasco, D., 53
Central Mexico, 17–55
Chavez, Gabriel, de, 31, 44, 62
Chilam Balam of Chumayel, 63,
    65–66
Chilam Balam of Kaua, 61–63
Christian confession, 24
Chumayel Wheel, 65–66
Circle, defined, 13
Circular dogma, 11–12
Clandestine documents, 8
Codex Aubin, 78
Codex Borbonicus, 73
Codex Mexicanus, 4, 6, 47–49
Codex Tlaxcala, 23–24, 30–31
Coiled snake imagery, 30–31
Compass, wind, 61–62, 89
Confession, 6, 24
Copernican revolution, 11, 15
Copernicus, 11–12, 45, 87
Crescents, 28, 30, 32–33, 64
Cross, pecked, 83–84
Cuesta, Domingo M., 19
Curers, 6

D

De Natura Rerum, 13. *See also* Liber
    Rotarum
De Proprietatibus, 45
De Revolutionibus, 11, 15, 45
Definition of cycle, 13
Devil, pact with, 8
Dias intercalares, 31
Dibble, G., 53
Dogma, 11–12

Dominical Letter, 23, 48
D'Olwer, L., 24
Dresden Codex, 73
Durán Calendar Wheel, 49–51
Dyer, N., 23

E
Earth-centered solar system, 11
Epicycles, 11
Eudoxus, 11
European Wheel, 21, 45

F
Florentine Codex, 41, 45

G
Garcia Cook, A. 4
Gemelli-Careri, F., 26
Gemelli-Careri Wheel, 29, 36
Gemelli Wheel, 30, 33, 54
Geometric modeling, 11
Gingerich, O., 16
Glass, J., 26, 28, 32, 36, 38, 41, 53,
    59, 78
Greeks, 11, 61
Gregorian Calendar Reform, 16, 47

H
Hecataeus, 87
Heritage, 8, 16
Historic time vs. mythic time, 53
Horcasitas, F., 49
Horoscopic system, 61

I
Icazbalceta, Garcia, 45
Identity, 8

Idolatry, 8, 46–47
Itzolkin, 3

J
Jansen, M., 81
Judeo-Christian concepts, 23

K
Katun Wheel, 59–61, 68
Kepler, Johannes, 12–13
Koestler, Arthur, 11
Kristeller, P., 11
Kubler, G., 26, 50, 53

L
Landa's Katun Wheel, 59–61
Liber Rotarum, 13
Lorenzana Wheel, 35

M
Madrid Codex, 73–74
Mapa de Cuauhtinchan, 53
Mappae mundi, 87
Mars, 12
Mathes, M., 45
Maya Calendar Wheels, 57–69
Mendoza Codex, 79
Mesoamerican time, 4, 33, 47, 50,
    68, 82, 87, 89
Milbrath, S., 61, 74
Miram, H., 45, 61, 68, 87
Mixtec–Puebla document, 74
Motolinía Wheel, 19–40, 46–47, 54
    Sahagún's refutation of, 46–47
Motolinía's History of the Indians of
    New Spain, 19
Mundy, B., 78

Muñoz Camargo, F., 23. *See also*
Codex Tlaxcala
Muñoz Camargo Wheels, 24, 33, 36
Mythic time *vs.* historic time, 53

N
Nahuals, 8
Nine Lords of Night, 4
Noguera, E., 4
North Star, 38

O
Old Testament patriarch taxonomy,
13
Olympiads, 46–47, 80
Omens, 8
Oudijk, M., 8, 19, 30, 33

P
Pact with devil, 8
Pardon of sins, 24
Patriarch taxonomy, Old Testament,
13
Peabody Wheel, 36
Pecked crosses, 83–84
Perujo, J., 30
Plato, 3, 11–12, 46
Principio xihuitladd, 19
Prognostication, 8
Ptolemy, 11

Q
Quadripartite universe, 88
Quintessence, 11

R
Ramirez, J.F., 25
Relaciones Geograficas de Tlaxcala,
23. *See also* Codex Tlaxcala

Religious identity, 8
Robertson, D., 45
Rotae, 12–13, 15, 54, 87
Roys, R., 64
Russo, A., 88

S
Sacred heritage, 8
Sahagún, B., de, 28, 41–47
refutation of Motolinía, 46–47
San Matias Thalancaleca, Puebla, 4
Santa Cruz de Tlateloco, 45
Serna, J., de la, 41–42
Seville, Isidore of, 13
Sin, 6, 24
Snake imagery, 30–31
Solar declinations, 19, 21
Sosa, J., 73
Spitler, S., 8, 23, 45, 50, 78, 80, 87
St. Andrew's Cross, 74
St. Peter, 47
Steck, F., 19, 23
Stuart, G., 3
Swastika image, 49–50, 74
Syllogisms, 13

T
Taube, K., 68
Tecpatl, 23
Tecpatl xihuitl, 19, 23, 26, 36, 78,
80
Tedlock, D., 73
Teeple, J.E., 3
Tenochtítlan, 53, 76
Teotihuacan pecked cross, 84
Texcoco, 26, 50, 53
Thompson, J.E.S., 68

Time-machine metaphor, 3–4
Tlateloco Manuscript, 41
Tochtli, 19, 23, 25–26, 53, 76, 78, 80
Tocqueville, Alexis, de, 19
Tonalpohualli, 4, 19, 41, 48, 74, 88
Tovar Calendar, 49–50
Tozzer, A., 4, 60, 68
Tzolkin, 67, 88

U
Uazlazon Katun, 61

V
Venerable Bede, 13
Venus glyphs, 61

Veytia Wheels, 34, 36–40, 78, 80
Villela, K., 68

W
Weeks, J., 8
Wheel of Time for Common Folk, 14
Wilkerson, J., 25
Wind compass, 61–62, 89

X
Xihuitl, 19, 54
Xiuhmolpilli, 48
spatial distribution, 80

Z
Zamorano, R., 8
Zodiac, 12, 45, 50, 61, 87, 89

www.ingramcontent.com/pod-product-compliance
Lightning Source LLC
Chambersburg PA
CBHW080920100426
42812CB00007B/2329